FM 23-35

**AUTOMATIC PISTOL
CALIBER .45
M1911 AND M1911A1**

FIELD MANUAL

BY WAR DEPARTMENT

©2013 Periscope Film LLC
All Rights Reserved
ISBN#978-1-940453-04-0
www.PeriscopeFilm.com

DISCLAIMER:

This manual is sold for historic research purposes only, as an entertainment. It contains obsolete information and is not intended to be used as part of an actual operation or maintenance training program. No book can substitute for proper training by an authorized instructor.

©2013 Periscope Film LLC
All Rights Reserved
ISBN#978-1-940453-04-0
www.PeriscopeFilm.com

FM 23-35

BASIC FIELD MANUAL

AUTOMATIC PISTOL, CALIBER .45
M1911 AND M1911A1

Prepared under direction of the
Chief of Cavalry

UNITED STATES
GOVERNMENT PRINTING OFFICE
WASHINGTON: 1940

For sale by the Superintendent of Documents, Washington, D. C. - Price 15 cents

WAR DEPARTMENT,
WASHINGTON, *April 30, 1940.*

FM 23–35, Automatic Pistol, Caliber .45, M1911 and 1911A1, is published for the information and guidance of all concerned.

[A. G. 062.11 (3–1–40).]

BY ORDER OF THE SECRETARY OF WAR:

G. C. MARSHALL,
Chief of Staff.

OFFICIAL:
E. S. ADAMS,
*Major General,
The Adjutant General.*

TABLE OF CONTENTS

		Paragraphs	Page
CHAPTER 1. MECHANICAL TRAINING.			
SECTION	I. Description	1– 2	1–5
	II. Disassembling and assembling	3– 4	5–8
	III. Care and cleaning	5– 11	8–11
	IV. Functioning	12– 14	11–14
	V. Spare parts and accessories	15– 16	14–15
	VI. Ammunition	17– 24	15–18
	VII. Individual safety precautions	25– 26	18–20
CHAPTER 2. MANUAL OF THE PISTOL, LOADING AND FIRING, DISMOUNTED AND MOUNTED.			
SECTION	I. General	27	21
	II. Dismounted	28– 37	22–24
	III. Mounted	38– 44	24–25
CHAPTER 3. MARKSMANSHIP, KNOWN-DISTANCE TARGETS, DISMOUNTED.			
SECTION	I. Preparatory training	45– 51	26–50
	II. Courses to be fired	52– 54	50–52
	III. Conduct of range practice	55– 61	52–61
	IV. Known-distance targets and ranges; range precautions	62– 64	61–65
	V. Small-bore practice	65– 69	65–66
CHAPTER 4. MARKSMANSHIP, KNOWN-DISTANCE TARGETS, MOUNTED.			
SECTION	I. Preparatory training	70– 72	67–73
	II. Courses to be fired	73– 74	73
	III. Conduct of range practice	75– 78	74–79
	IV. Known-distance targets and ranges; range precautions	79– 81	79–80
	V. Small-bore practice	82– 84	80
CHAPTER 5. FIRING AT FIELD TARGETS.			
SECTION	I. Dismounted	85– 88	81–82
	II. Mounted	89– 93	82–85
CHAPTER 6. ADVICE TO INSTRUCTORS.			
SECTION	I. General	94– 95	86–87
	II. Mechanical training	96–101	88–89
	III. Manual of the pistol	102	89
	IV. Marksmanship	103–106	90–92

FM 23-35

BASIC FIELD MANUAL

AUTOMATIC PISTOL, CALIBER .45, M1911 AND M1911A1

(The matter contained herein supersedes TR 1300-45A, January 3, 1938, and chapter 3, part one, Basic Field Manual, volume III, April 5, 1932.)

CHAPTER 1

MECHANICAL TRAINING

	Paragraphs
SECTION I. Description	1-2
II. Disassembling and assembling	3-4
III. Care and cleaning	5-11
IV. Functioning	12-14
V. Spare parts and accessories	15-16
VI. Ammunition	17-24
VII. Individual safety precautions	25-26

SECTION I

DESCRIPTION

■ 1. GENERAL.—*a.* The automatic pistols, caliber .45, M1911 and M1911A1, are recoil-operated, magazine-fed, self-loading hand weapons (figs. 1 and 2). The gas generated in a cartridge fired in the pistol is utilized to perform the functions of extracting and ejecting the empty cartridge case, cocking the hammer and forcing the slide to the rearmost position, thereby compressing the recoil spring. The action of the recoil spring forces the slide forward, feeding a live cartridge from the magazine into the chamber, leaving the weapon ready to fire again.

b. The M1911A1 pistol is a modification of the M1911 pistol. The operation of both models of pistols is exactly the same. The changes consist of the following (fig. 2):

(1) The tang of the grip safety is extended better to protect the hand.

(2) A clearance cut is made on the receiver for the trigger finger.

1

(3) The face of the trigger is cut back and knurled.

(4) The mainspring housing is raised in the form of a curve to fit the palm of the hand and is knurled.

(5) The top of the front sight is widened.

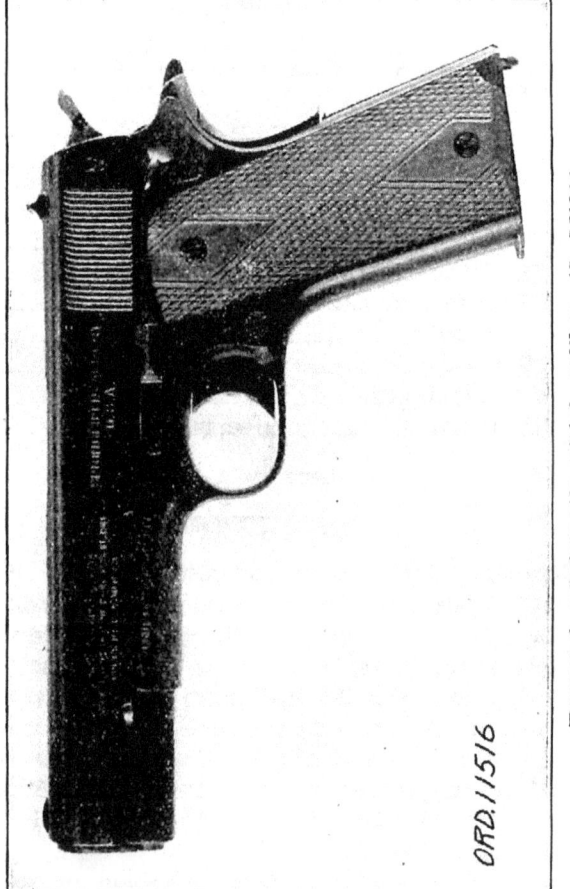

Figure 1.—Automatic pistol, caliber .45, M1911.

c. The pistol is designed to fire cartridge, ball, caliber .45, M1911. The magazine holds seven cartridges. The upper cartridge is stripped from the magazine and forced into the chamber by the forward motion of the slide. The pistol fires but once at each squeeze of the trigger. When the last cartridge in the magazine has been fired the slide remains open. The magazine catch is then depressed and the empty magazine falls out. A loaded magazine is then inserted, making seven more shots available.

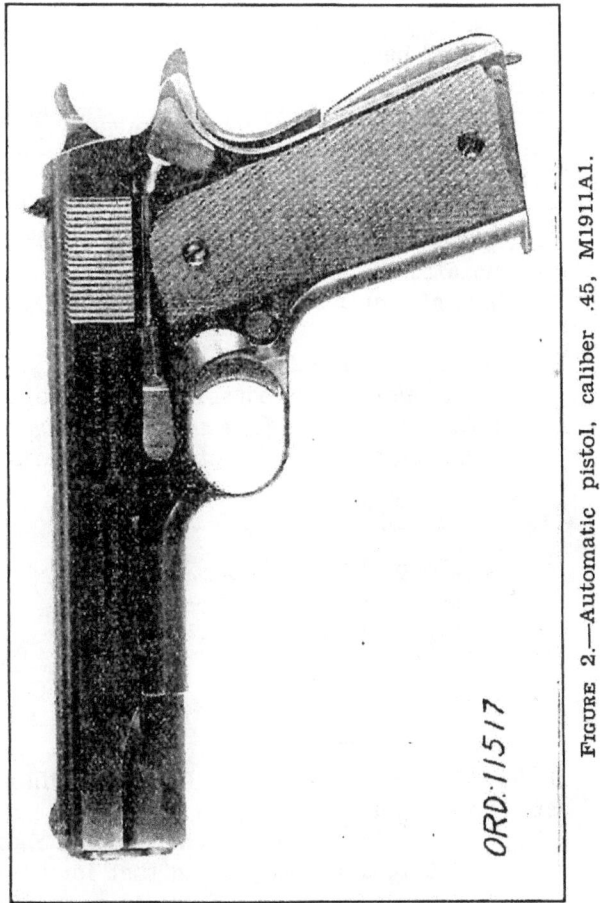

FIGURE 2.—Automatic pistol, caliber .45, M1911A1.

d. The rate of fire is limited by the dexterity of the operator in inserting magazines into the pistol and the ability of the firer to aim and squeeze.

2. GENERAL DATA.—a. *Dimensions.*

(1) *Barrel.*

Caliber of bore _____inches__	0.45
Number of grooves _____	6
Twist in rifling, uniform L. H., one turn in inches__	16
Length of barrel _____do__	5.03

(2) *Pistol.*

Overall length of pistol _____inches__	8.593
Height of front sight above axis of bore__do__	0.5597

b. *Weights.*

Weight of pistol with magazine_____pounds__	2.437
Weight of loaded magazine, 7 rounds (approximate) _____do__	0.481
Weight of empty magazine_____do__	0.156

c. *Trigger pull.*

Pistols, new or repaired_____pounds__	5½ to 6½
Pistols, in hands of troops_____do__	5 to 6½

d. *Exterior ballistics.*—(1) *Accuracy with muzzle rest.*—The figures indicated below represent the mean variations for several targets.

Range	Mean radius	Mean vertical deviation
Yards	*Inches*	*Inches*
25	0.86	0.62
50	1.36	0.91
75	2.24	1.42

(2) *Drift.*—The drift or deviation due to the rifling in this pistol is to the left, but is more than neutralized by the pull of the trigger when the pistol is fired from the right hand. The drift is slight at short ranges and that for long ranges is immaterial, inasmuch as the pistol is a comparatively short-range weapon.

AUTOMATIC PISTOL, CAL. .45, M1911 AND M1911A1   2–3

(3) *Velocity with striking energy.*

Range	Velocity	Energy	Range	Velocity	Energy
Yards	Feet per second	Foot-pounds	Yards	Feet per second	Foot-pounds
0	802	329	150	717	262
25	788	317	175	704	253
50	773	305	200	691	244
75	758	294	225	678	235
100	744	283	250	666	226
125	730	272			

(4) *Penetration in white pine.*

Range	Depth	Range	Depth
Yards	Inches	Yards	Inches
25	6.0	150	5.2
50	5.8	200	4.6
75	5.6	250	4.0
100	5.5		

A penetration of 1 inch in white pine corresponds to a dangerous wound. The penetration in moist loam at 25 yards is about 10 inches. The penetration in dry sand at 25 yards is about 8 inches.

(5) *Trajectory.*—The elevation required for 100 yards is 24' and for 200 yards about 1°. The elevation for 250 yards is about 1° 13'; the maximum ordinate being approximately 130 yards distance from the muzzle and about 51 inches in height. The maximum range is approximately 1,600 yards at an angle of elevation of 30°. The maximum ordinate for the maximum range is approximately 2,000 feet.

Section II

DISASSEMBLING AND ASSEMBLING

■ 3. DISASSEMBLING (see fig. 3).—*a.* Remove the magazine by pressing the magazine catch.

b. Press the recoil spring plug inward and turn the barrel bushing to the right until the recoil spring plug and the end of the recoil spring protrude from their seat, releasing the tension of the recoil spring. As the recoil spring plug is allowed to protrude from its seat, the finger or thumb should be kept over it so that it will not jump away and be lost or strike the operator. Draw the slide rearward until the smaller rear recess in its lower left edge stands above the projection on the thumbpiece of the slide stop; press gently against the end of the pin of the slide stop which protrudes from the right side of the receiver above the trigger guard and remove the slide stop.

c. This releases the barrel link, allowing the barrel with the barrel link and the slide to be drawn forward together from the receiver, carrying with them the barrel bushing, recoil spring, recoil spring plug, and recoil spring guide.

d. Remove these parts from the slide by withdrawing the recoil spring guide from the rear of the recoil spring and drawing the recoil spring plug and the recoil spring forward from the slide. Turn recoil spring plug to right to remove from recoil spring. Turn the barrel bushing to the left until it may be drawn forward from the slide. This releases the barrel which with the barrel link may be drawn forward from the slide, and by pushing out the barrel link pin the barrel link is released from the barrel.

e. Press the rear end of the firing pin forward until it clears the firing pin stop which is then drawn downward from its seat in the slide; the firing pin, firing pin spring, and extractor are then removed from the rear of the slide.

f. The safety lock is readily withdrawn from the receiver by cocking the hammer and pushing from the right on the pin part or pulling outward on the thumbpiece of the safety lock when it is midway between its upper and lower positions. The cocked hammer is then lowered and removed after removing the hammer pin from the left side of the receiver. The mainspring housing pin is then pushed out from the right side of the receiver which allows the mainspring housing to be withdrawn downward and the grip safety rearward from the handle. The sear spring may then be removed. By pushing out the sear pin from the right to the left side of the receiver, the sear and the disconnector are released.

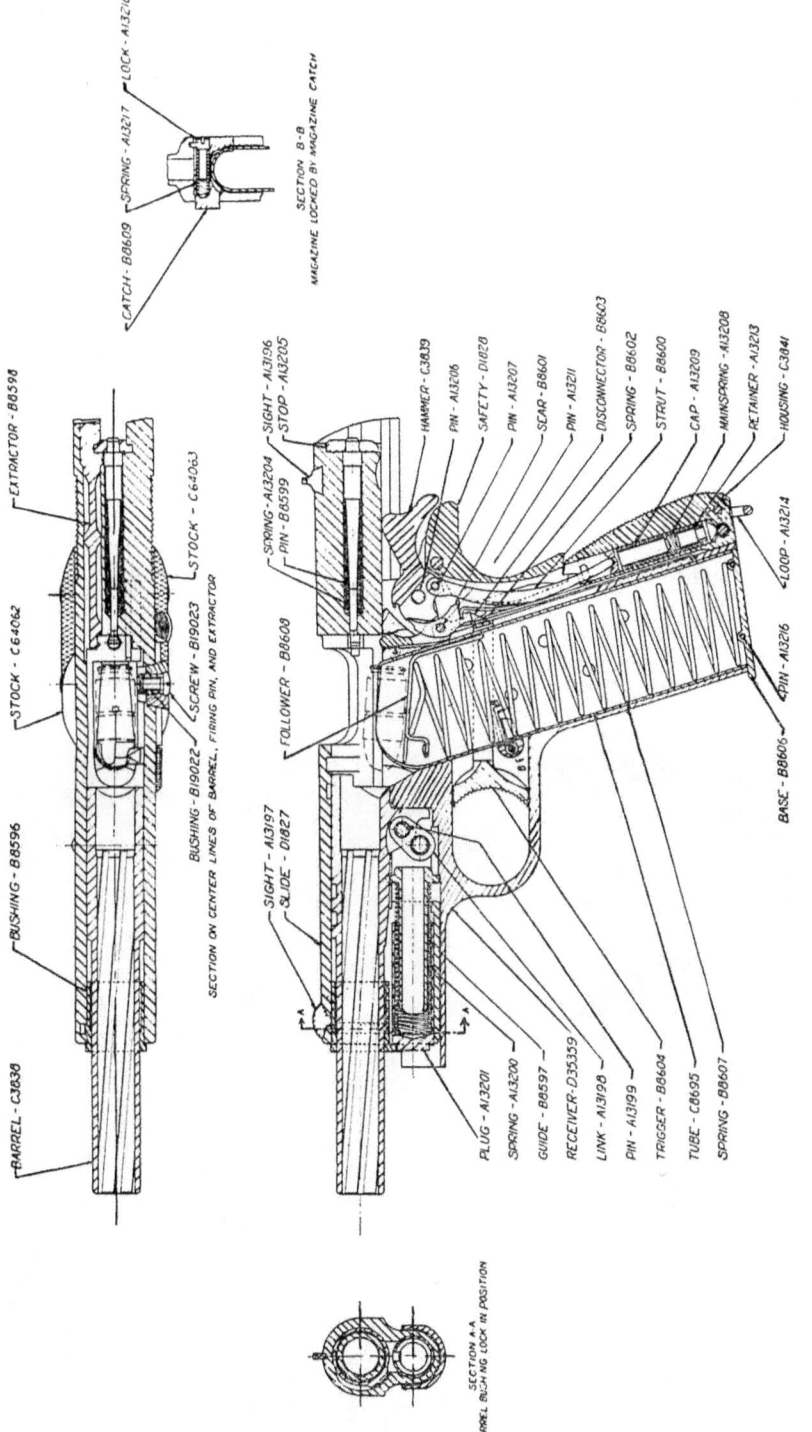

FIGURE 3.—Automatic pistol, caliber .45, M1911A1, sectional views.

AUTOMATIC PISTOL, CAL. .45, M1911 AND M1911A1   3–4

g. To remove the mainspring, mainspring cap, and housing pin retainer from the mainspring housing, compress the mainspring and push out the small mainspring cap pin.

h. To remove the magazine catch from the receiver, its checkered left end must be pressed inward, when the right end of the magazine catch will project so far from the right side of the receiver that it may be rotated one-half turn. This movement will release the magazine-catch lock from its seat in the receiver, when the magazine catch, the magazine catch lock, and the magazine catch spring may be removed.

i. With the improved design of magazine catch lock the operation of dismounting the magazine catch is simplified. When the magazine catch has been pressed inward the magazine catch lock is turned a quarter turn to the left by means of a screw driver, or the short leaf of the sear spring. The magazine catch with its contents can then be removed. The improved design will be recognized from the fact that the head of the magazine catch lock is slotted.

j. The trigger can then be removed rearward from the receiver.

k. The hammer strut or the long arm of the screw driver can be used to push out all the pins except the mainspring cap pin, lanyard loop pin, and ejector pin.

l. The slide stop plunger, the safety lock plunger, and the plunger spring may be pushed to the rear out of the plunger tube.

m. The magazine should not be disassembled except for cleaning or to replace the magazine follower or magazine spring. To disassemble proceed as follows: Push the magazine follower downward about one-fourth inch; this compresses the magazine spring. Insert the end of a drift through one of the small holes in the side of the magazine to hold the magazine spring, then slide out the magazine follower. Hold hand over end of the magazine before removing drift from hole to prevent magazine spring from jumping away.

■ 4. ASSEMBLING.—*a.* Proceed in the reverse order.

b. It should be noted that the disconnector and sear are assembled as follows: Place the cylindrical part of the disconnector in its hole in the receiver with the flat face of the lower part of the disconnector resting against the yoke of the trigger. Then place the sear, lugs downward, so that it straddles the disconnector. The sear pin is then inserted in place so that it passes through both the disconnector and the sear.

c. The sear, disconnector, and hammer being in place and the hammer down, to replace the sear spring, locate its lower end in the cut in the receiver with the end of the long leaf resting on the sear; then insert the mainspring housing until its lower end projects below the frame about one-eighth inch, replace the grip safety, cock the hammer, and replace the safety lock; then lower the cocked hammer, push the mainspring housing home, and insert the mainspring housing pin.

d. In assembling the safety lock to the receiver use the tip of the magazine follower or the screw driver to press the safety lock plunger home, thus allowing the seating of the safety lock. It should be remembered that when assembling the safety lock the hammer must be cocked.

e. When replacing the slide and barrel on the receiver care must be taken that the barrel link is tilted forward as far as possible and that the barrel link pin is in place.

SECTION III

CARE AND CLEANING

■ 5. GENERAL.—*a.* Careful, conscientious work is required to keep automatic pistols in a condition that will insure perfect functioning of the mechanism and continued accuracy of the barrel. It is essential that the entire mechanism is kept cleaned and oiled to avoid jams.

b. The mechanism also requires care to prevent rust or an accumulation of sand or dirt in the interior. Pistols are easily disassembled for cleaning and oiling.

■ 6. CARE AND CLEANING.—*a.* Care and cleaning of the pistol include the ordinary case of the pistol to preserve its condition and appearance in garrisons, posts, and camps, and in campaign.

b. Damp air and sweaty hands are great promoters of rust. The pistols should be cleaned and protected after every drill or handling. Special precautions are necessary when the pistols have been used on rainy days and after tours of guard duty.

c. To clean the pistol rub it with a rag which has been lightly oiled and then clean with a perfectly dry rag. Swab the bore with an oily flannel patch and then with a perfectly dry one. Dust out all crevices with a small, clean brush.

d. Immediately after cleaning, to protect the pistol swab the bore thoroughly with a flannel patch saturated with sperm oil, wipe over all metal parts with an oily rag, applying a few drops of light oil (sperm oil) to all cams and working surfaces of the mechanism.

e. After cleaning and protecting the pistol, place it in the pistol rack without any covering whatever. The use of canvas or similar covers is prohibited, as they collect moisture and rust the metal parts. While barracks are being swept, pistol racks will be covered with a piece of canvas to protect the pistols from dust.

■ 7. CARE AND CLEANING AFTER FIRING.—*a.* When a pistol has been fired the bore will be cleaned thoroughly not later than the evening of the day on which it is fired. Thereafter it will be cleaned and oiled each day for at least the next three succeeding days.

b. To clean the bore after firing, first remove the slide and barrel, insert the muzzle of the barrel in a vessel containing hot water and issue soap, hot water alone, or cold water; the cleaning rod with a cloth patch assembled is inserted in the breech and moved forward and back for about 1 minute, pumping the water in and out of the bore. When the bore is wet, a brass or bronze wire brush, if available, should be run all the way through the bore, then all the way back three or four times. Water should again be pumped through the bore. Then wipe the cleaning rod dry, remove the barrel from the water, and using dry, clean flannel patches thoroughly swab the bore until it is perfectly dry and clean. Examine the bore carefully for metal fouling.

7-11 AUTOMATIC PISTOL, CAL. .45, M1911 AND M1911A1

CAUTION.—After firing do not oil the bore before cleaning.

  c. Saturate a clean flannel patch with sperm oil and swab the bore and chamber with the patch, making certain that the bore and all metal parts of the pistol are covered with a thin coat of oil.

■ 8. Rules for Care of Pistol on the Range.—*a.* Always clean at the end of each day's shooting. A pistol that has been fired should not be left over night without cleaning.

  b. Never fire a pistol with any dust, dirt, mud, or snow in the bore.

  c. Before loading the pistol make sure that no patch, rag, or other object has been left in the barrel.

  d. During range firing a noncommissioned officer will be placed in charge of the cleaning of pistols in the cleaning racks.

■ 9. Care During Cold Weather.—Use oil sparingly on the working parts.

■ 10. Care During Gas Attacks.—*a.* Pistols should be cleaned as soon as possible after a gas attack.

  b. Oil will prevent corrosion for about 12 hours.

  c. Clean all parts in boiling water containing a little soda, if available.

  d. All traces of gas must be removed from ammunition with a slightly oiled rag; then thoroughly dry the ammunition

  e. Rust-preventive compound resists gas corrosion more than light oil. In many exposures, especially those of long duration, ammunition treated with sperm oil evidences more severe corrosion than unprotected cartridges.

■ 11. Important Points To Be Observed.—*a.* After firing the pistol, never leave it uncleaned over night. The damage done is then irreparable.

  b. Keep the pistol clean and lightly lubricated, but do not let it become gummy with oil.

  c. Do not place the pistol on the ground where sand or dirt may enter the bore or mechanism.

  d. Do not plug the muzzle of the pistol with a patch or plug. One may forget to remove it before firing, in which case the discharge may bulge or burst the barrel at the muzzle.

  e. A pistol kept in a leather holster may rust due to moisture absorbed by the leather from the atmosphere, even though the holster may appear to be perfectly dry. If the holster is wet and the pistol must be carried therein, cover the pistol with a thick coat of oil.

f. The hammer should not be snapped when the pistol is partially disassembled.

g. The trigger should be pulled with the forefinger. If the trigger is pulled with the second finger, the forefinger extending along the side of the receiver is apt to press against the projecting pin of the slide stop and cause a malfunction when the slide recoils.

h. Pressure on the trigger must be released sufficiently after each shot to permit the trigger to reengage the sear.

i. To remove cartridges not fired, disengage the magazine slightly and then extract the cartridge in the barrel by drawing back the slide.

j. Care should be taken to see that the magazine is not dented or otherwise damaged.

k. Care must be exercised in inserting the magazine to insure its engaging with the magazine catch. Never insert the magazine and strike it smartly with the hand to force it home, as this may spring the base or the inturning lips at the top. It should be inserted by a quick continuous movement.

SECTION IV

FUNCTIONING

■ 12. METHOD OF OPERATION.—*a.* A loaded magazine is placed in the receiver and the slide drawn fully back and released, thus bringing the first cartridge into the chamber. (If the slide is open push down the slide stop to let the slide go forward.) The hammer is thus cocked and the pistol is ready for firing.

b. If it is desired to make the pistol ready for instant use and for firing the maximum number of shots with the least possible delay, draw back the slide, insert a cartridge by hand into the chamber of the barrel, allow the slide to close, then lock the slide and the cocked hammer by pressing the safety lock upward and insert a loaded magazine. The slide and hammer being thus positively locked, the pistol may be carried safely at full cock and it is only necessary to press down the safety lock (which is located within easy reach of the thumb) when raising the pistol to the firing position.

c. The grip safety is provided with an extending horn which not only serves as a guard to prevent the hand of the shooter from slipping upward and being struck or injured by the hammer, but also aids in accurate shooting by keeping the hand in the same position for each shot and, furthermore, permits the lowering of the cocked hammer with one hand by automatically pressing in the grip safety when the hammer is drawn slightly beyond the cocked position. In order to release the hammer, the grip safety must be pressed in before the trigger is squeezed.

■ 13. SAFETY DEVICES.—*a.* It is impossible for the firing pin to discharge or even touch the primer except on receiving the full blow of the hammer.

b. The pistol is provided with two automatic safety devices:

(1) The disconnector, which positively prevents the release of the hammer unless the slide and barrel are in the forward position and safely interlocked. This device also controls the firing and prevents more than one shot from following each squeeze of the trigger.

(2) The grip safety which at all times locks the trigger unless the handle is firmly grasped and the grip safety pressed in.

c. In addition, the pistol is provided with a safety lock by which the closed slide and the cocked hammer can be positively locked in position.

■ 14. DETAILED FUNCTIONING.—*a.* The magazine may be charged with any number of cartridges from one to seven.

b. The charged magazine is inserted in the receiver and the slide drawn once to the rear. This movement cocks the hammer, compresses the recoil spring, and when the slide reaches the rear position the magazine follower raises the upper cartridge into the path of the slide. The slide is then released and being forced forward by the recoil spring carries the first cartridge into the chamber of the barrel. As the slide approaches its forward position, it encounters the rear extension of the barrel and forces the barrel forward; the rear end of the barrel swings upward on the barrel link, turning on the muzzle end as on a fulcrum. When the slide and barrel reach their forward position they are positively locked together by the locking ribs on the barrel and their joint forward movement is arrested by the barrel lug encountering the pin on the slide top. The pistol is then ready for firing.

c. When the hammer is cocked the hammer strut moves downward, compressing the mainspring, and the sear under action of the long leaf of the sear spring engages its nose in the notch on the hammer. In order that the pistol may be fired the following conditions must exist:

(1) The grip safety must be pressed in, leaving the trigger free to move.

(2) The slide must be in its forward position, properly interlocked with the barrel so that the disconnector is held in the recess on the under side of the slide under the action of the sear spring, transmitting in this position any motion of the trigger to the sear.

(3) The safety lock must be down in the unlocked position so that the sear will be unblocked and free to release the hammer and the slide will be free to move back.

d. On squeezing the trigger, the sear is moved and the released hammer strikes the firing pin which transmits the blow to the primer of the cartridge. The pressure of the gases generated in the barrel by the explosion of the powder in the cartridge is exerted in a forward direction against the bullet, driving it through the bore, and in a rearward direction against the face of the slide, driving the latter and the barrel to the rear together. The downward swinging movement of the barrel unlocks it from the slide and the barrel is then stopped in its lowest position. The slide continues to move to the rear, opening the breech, cocking the hammer, extracting and ejecting the empty shell, and compressing the recoil spring until the slide reaches its rearmost position when another cartridge is raised in front of it and forced into the chamber of the barrel by the return movement of the slide under pressure of the recoil spring.

e. The weight and consequently the inertia of the slide augmented by those of the barrel are so many times greater than the weight and inertia of the bullet that the latter has been given its maximum velocity and has been driven from the muzzle of the barrel before the slide and barrel have recoiled to the point where the barrel commences its unlocking movement. This construction therefore delays the opening of the breech of the barrel until after the bullet has left the muzzle and therefore practically prevents the escape of any of the powder gases to the rear after the breech has been opened. This factor of safety is further increased by the tension of the recoil spring and mainspring, both of which oppose the rearward movement of the slide.

f. While the comparatively great weight of the slide of the pistol insures safety against premature opening of the breech, it also insures operation of the pistol because at the point of the rearward opening movement where the barrel is unlocked and stopped the heavy slide has attained a momentum which is sufficient to carry it through its complete opening movement and makes the pistol ready for another shot.

g. When the magazine has been emptied, the pawl-shaped slide stop is raised by the magazine follower under action of the magazine spring into the front recess on the lower left side of the slide, thereby locking the slide in the open position and serving as an indicator to remind the shooter that the empty magazine must be replaced by a loaded one before the firing can be continued. Pressure upon the magazine catch quickly releases the empty magazine from the receiver and permits the insertion of a loaded magazine.

h. To release the slide from the open position, it is only necessary to press upon the thumbpiece of the slide stop, then the slide will go forward to its closed position, carrying a cartridge from the previously inserted magazine into the barrel and making the pistol ready for firing again.

SECTION V

SPARE PARTS AND ACCESSORIES

■ 15. SPARE PARTS.—In time certain parts of the pistol become unserviceable through breakage or wear resulting from continuous usage. For this reason spare parts are provided for replacement purposes. They should be kept clean and lightly oiled to prevent rust. They are divided into two groups; spare parts and basic spare parts.

a. Spare parts.—These are extra parts provided with the pistol for replacement of the parts most likely to fail, for use in making minor repairs, and in general care of the pistol. Sets of spare parts should be kept complete at all times. Whenever a spare part is taken to replace a defective part in the pistol, the defective part should be repaired or a new one substituted in the spare part set as soon as possible. The allowance of these spare parts is prescribed in SNL B-6.

b. Basic spare parts.—These are sets of parts provided for the use of ordnance maintenance companies and include all parts necessary to repair the pistol. The allowance of basic spare parts is prescribed in the addendum to SNL B-6.

■ 16. ACCESSORIES.—The names or general characteristics of many of the accessories required with the automatic pistol indicate their use and application. They consist of the holster, lanyard, and pistol cleaning kit, and for post, camp, or station issue, arm lockers and arm racks. The pistol cleaning kit contains cleaning brushes and rods, pistol screwdrivers, an oiler, and a small brass can in which the set of spare parts is carried.

SECTION VI

AMMUNITION

■ 17. GENERAL.—The information in this section pertaining to the ammunition authorized for use in the automatic pistol, cal. .45, M1911 and M1911A1, includes a description of the cartridges, means of identification, care, use, and ballistic data.

■ 18. CLASSIFICATION.—The types of ammunition provided for this pistol are—
 a. Ball, for use against personnel and light matériel targets.
 b. Dummy, for training (cartridges are inert).

■ 19. AMMUNITION LOT NUMBER.—When ammunition is manufactured an ammunition lot number which becomes an essential part of the marking is assigned in accordance with pertinent specifications. This lot number is marked on all packing containers and on the identification card inclosed in each packing box. It is required for all purposes of record, to handle. However, care must be observed to keep the boxes from becoming broken or damaged. All broken boxes must be immediately repaired and careful attention given so that all markings are transferred to the new parts of the box. The metal liner should be air-tested and sealed if equipment for this work is available.

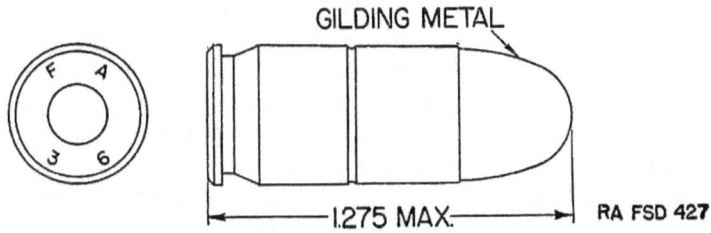

Figure 4.—Cartridge, ball, cal. .45, M1911.

b. Ammunition boxes should not be opened until the ammunition is required for use. Ammunition removed from the airtight container, particularly in damp climates, is apt to corrode, thereby causing the ammunition to become unserviceable.

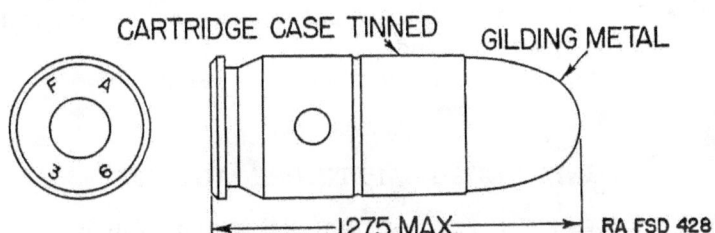

Figure 5.—Cartridge, dummy, cal. .45, M1921.

c. Carefully protect the ammunition from mud, sand, dirt, and water. If it gets wet or dirty wipe it off at once. If veridigris or light corrosion forms on cartridges, it should be wiped off. However, cartridges should not be polished to make them look better or brighter.

d. The use of oil or grease on cartridges is dangerous and is prohibited.

e. Do not fire dented cartridges, cartridges with loose bullets, or otherwise defective rounds.

22-25 AUTOMATIC PISTOL, CAL. .45, M1911 AND M1911A1

f. Do not allow the ammunition to be exposed to the direct rays of the sun for any length of time. This is liable to affect seriously its firing qualities.

g. No caliber .45 ammunition will be fired until it has been positively identified by ammunition lot number and grade as published in the latest revision or change to Ordnance Field Service Bulletin No. 3-5.

■ 23. STORAGE.—*a.* Whenever practicable small-arms ammunition should be stored under cover. Should it become necessary to leave small-arms ammunition in the open it should be raised on dunnage at least 6 inches from the ground and the pile covered with a double thickness of paulin. Suitable trenches should be dug to prevent water flowing under the pile.

b. Fire hazard.—If fired into or placed in a fire, small-arms ammunition does not explode violently. There are small individual explosions of each cartridge, the case flying in one direction and the bullet in another. In case of fire it is advisable to keep those not engaged in fighting the fire at least 200 yards from the fire and have them lie on the ground. It is unlikely that the bullets and cases will fly over 200 yards.

■ 24. BALLISTIC DATA.—*a.* Average velocity at 25 feet from muzzle, 800 feet per second.

b. Approximate maximum range, 1,600 yards.

SECTION VII

INDIVIDUAL SAFETY PRECAUTIONS

■ 25. RULES FOR SAFETY.—Before ball ammunition is issued, the soldier must know the essential rules for safety with the pistol. The following rules are taught as soon as the recruit is sufficiently familiar with the pistol to understand them. They should be enforced by constant repetition and coaching until their observance becomes the soldier's fixed habit when handling the pistol. When units carrying the pistol are first formed, the officer or noncommissioned officer in charge causes the men to execute INSPECTION PISTOL.

a. Execute UNLOAD every time the pistol is picked up for any purpose. Never trust your memory. Consider every pistol as loaded until you have proved it otherwise.

18

AUTOMATIC PISTOL, CAL. .45, M1911 AND M1911A1   25

b. Always unload the pistol if it is to be left where someone else may handle it.

c. Always point the pistol up when snapping it after examination. Keep the hammer fully down when the pistol is not loaded.

d. Never place the finger within the trigger guard until you intend to fire or to snap for practice.

e. Never point the pistol at anyone you do not intend to shoot, nor in a direction where an accidental discharge may do harm. On the range, do not snap for practice while standing back of the firing line.

f. Before loading the pistol, draw back the slide and look through the bore to see that it is free from obstruction.

g. On the range, do not insert a loaded magazine until the time for firing.

h. Never turn around at the firing point while you hold a loaded pistol in your hand, because by so doing you may point it at the man firing alongside of you.

i. On the range, do not load the pistol with a cartridge in the chamber until immediate use is anticipated. If there is any delay, lock the pistol and only unlock it while extending the arm to fire. Do not lower the hammer on a loaded cartridge; the pistol is much safer cocked and locked.

j. In reducing a jam *first remove the magazine.*

k. To remove a cartridge not fired *first remove the magazine* and then extract the cartridge from the chamber by drawing back the slide.

l. In campaign, when early use of the pistol is not foreseen, it should be carried with a fully loaded magazine in the socket, chamber empty, hammer down. When early use of the pistol is probable, it should be carried loaded and locked in the holster or hand. In campaign, extra magazines should be carried fully loaded.

m. When the pistol is carried in the holster loaded, cocked, and locked the butt should be rotated away from the body when drawing the pistol in order to avoid displacing the safety lock.

n. Safety devices should be frequently tested. A safety device is a dangerous device if it does not work when expected.

■ 26. TESTS.—*a. Safety lock.*—Cock the hammer and then press the safety lock upward into the safe position. Grasp the stock so that the grip safety is depressed and squeeze the trigger three or four times. If the hammer falls, the safety lock is not safe and must be repaired.

b. Grip safety.—Cock the hammer and, being careful not to depress the grip safety, point pistol downward and squeeze the trigger three or four times. If the hammer falls or the grip safety is depressed by its own weight, the grip safety is not safe and must be repaired.

c. Half-cock notch.—Draw back the hammer until the sear engages the half-cock notch and squeeze the trigger. If the hammer falls, the hammer or sear must be replaced or repaired. Draw the hammer back nearly to full cock and then let it slip. It should fall only to half cock.

d. Disconnector.—Cock the hammer. Shove the slide one-quarter inch to the rear; hold slide in that position and squeeze the trigger. Let the slide go forward, maintaining the pressure on the trigger. If the hammer falls, the disconnector is worn on top and must be replaced. Pull the slide all the way to the rear and engage the slide stop. Squeeze the trigger and at the same time release the slide. The hammer should not fall. Release the pressure on the trigger and then squeeze it. The hammer should then fall. The disconnector prevents the release of the hammer unless the slide and barrel are in the forward position safely interlocked. It also prevents more than one shot following each squeeze of the trigger.

CHAPTER 2

MANUAL OF THE PISTOL, LOADING AND FIRING, DISMOUNTED AND MOUNTED

	Paragraphs
SECTION I. General	27
II. Dismounted	28–37
III. Mounted	38–44

SECTION I

GENERAL

■ 27. GENERAL.—*a.* The movements herein described differ in purpose from the manual of arms for the rifle in that they are not designed to be executed in exact unison. Furthermore, with only a few exceptions, there is no real necessity for their simultaneous execution. They are not therefore planned as a disciplinary drill to be executed in cadence with snap and precision, but merely as simple, quick, and safe methods of handling the pistol. Commands are prescribed only for such movements as may be occasionally executed simultaneously by the squad or larger unit.

b. In general, movements begin and end at the position of RAISE PISTOL.

c. Commands for firing, when required, are limited to COMMENCE FIRING and CEASE FIRING.

d. Officers and enlisted men armed with the pistol remain at the position of ATTENTION during the manual of arms, except when their units are presented to their commanders or are presented during ceremonies, at retreat, and at guard mounting. In such cases they execute the hand salute at the command of execution ARMS of 1. PRESENT, 2. ARMS, and resume the position of attention at the command of execution of the next command.

e. The lanyard is used whenever the pistol is carried mounted. The lanyard should be of such length that the arm may be fully extended without constraint.

Section II

DISMOUNTED

■ 28. To RAISE PISTOL (fig. 6).—The commands are: 1. RAISE, 2. PISTOL. At the command PISTOL, unbutton the flap of the holster with the right hand and grasp the stock, back of the hand outward. Draw the pistol from the holster; reverse it, muzzle up, the thumb and last three fingers holding the stock, the forefinger extended outside the trigger guard, the barrel of the pistol to the rear and inclined to the front at an angle of 30°, the hand as high as, and 6 inches in front of, the point of the right shoulder. This is the position of RAISE PISTOL.

■ 29. To WITHDRAW THE MAGAZINE (fig. 6).—Without lowering the right hand, turn the barrel slightly to the right; press the magazine catch with the right thumb and with the left hand remove the magazine. Place it in the belt or pocket.

■ 30. To OPEN THE CHAMBER (fig. 6).—Withdraw the magazine and resume the position of RAISE PISTOL. Without lowering the right hand, grasp the slide with the thumb and the first two fingers of the left hand (thumb on left side of slide and pointing downward); keeping the muzzle elevated, shift the grip of the right hand so that the right thumb engages with the slide stop. Push the slide downward to its full extent and force the slide stop into its notch with the right thumb without lowering the muzzle of the pistol.

■ 31. To CLOSE THE CHAMBER.—With the right thumb press down the slide stop and let the slide go forward. Squeeze the trigger.

■ 32. To INSERT A MAGAZINE.—Without lowering the right hand, turn the barrel to the right. Grasp a magazine with the first two fingers and thumb of the left hand; withdraw it from the belt and insert it in the pistol. Press it fully home.

■ 33. To LOAD PISTOL.—The commands are: 1. LOAD, 2. PISTOL. At the command PISTOL, if a loaded magazine is not already in the pistol, insert one. Without lowering the right hand, turn the barrel slightly to the left. Grasp the slide with the thumb and fingers of the left hand (thumb on right side of slide and pointing upward). Pull the slide downward to its full extent (fig. 6). Release the slide and engage the safety lock.

■ 34. To UNLOAD PISTOL.—The commands are: 1. UNLOAD, 2. PISTOL. At the command PISTOL, withdraw the magazine. Open the chamber as prescribed in paragraph 30. Glance at the chamber to verify that it is empty. Close the chamber. Take the position of RAISE PISTOL and squeeze the trigger. Then insert an empty magazine.

① To raise the pistol. ② To withdraw the magazine. ③ To pull the slide downward in loading. ④ To open the chamber. ⑤ To inspect the pistol.

FIGURE 6.—Manual of the pistol (dismounted).

■ 35. To INSPECT PISTOL (fig. 6).—The commands are: 1. INSPECTION, 2. PISTOL. At the command PISTOL, withdraw the magazine. Open the chamber as prescribed in paragraph 30. Take the position of RAISE PISTOL. The withdrawn magazine is held in the open left hand at the height of the belt. After the pistol has been inspected, or at the command 1. RETURN, 2. PISTOL, close the chamber, take the position of RAISE PISTOL, and squeeze the trigger. Insert an empty magazine and execute RETURN PISTOL.

■ 36. To Return Pistol.—The commands are: 1. RETURN, 2. PISTOL. At the command PISTOL, lower the pistol to the holster, reversing it, muzzle down, back of the hand to the right; raise the flap of the holster with the right thumb; insert the pistol in the holster and thrust it home; button the flap of the holster with the right hand.

■ 37. To Fire the Pistol.—FULLY LOADED WITH BALL AMMUNITION.—Squeeze the trigger for each shot. When the last cartridge has been fired the slide will remain in the rear position with the chamber open.

SECTION III

MOUNTED

■ 38. General Rules.—The following movements are executed as when dismounted: RAISE PISTOL, RETURN PISTOL, CLOSE CHAMBER, TO FIRE THE PISTOL. The mounted movements may be practiced when dismounted by first cautioning, "Mounted position." The right foot is then carried 20 inches to the right and the left hand to the position of the bridle hand. Whenever the pistol is lowered into the bridle hand, the movement is executed by rotating the barrel to the right. Grasp the slide in the full grip of the left hand, thumb extending along the slide, back of the hand down, barrel down and pointing upward and to the left front.

■ 39. To Withdraw the Magazine.—Lower the pistol into the bridle hand. Press the magazine catch with the forefinger of the right hand, palm of the hand over the base of the magazine to prevent it from springing out; withdraw the magazine and place it in the belt or pocket.

■ 40. To Open the Chamber.—Withdraw the magazine. Grasp the stock with the right hand, back of the hand down, thrust forward and upward with the right hand, and engage the slide stop by pressure of the right thumb.

■ 41. To Insert a Magazine.—Lower the pistol into the bridle hand. Extra magazines should be carried in the belt with the projection on the base pointing to the left. Grasp the magazine with the tip of the right forefinger on the projection, withdraw it from the belt, and insert it in the pistol. Press it fully home.

■ 42. To Load Pistol.—The commands are: 1. LOAD, 2. PISTOL. At the command PISTOL, lower the pistol into the bridle hand. If a loaded magazine is not already in the pistol, insert one. Grasp the stock with the right hand, back of the hand down, and thrust upward and to the left front; release the slide and engage the safety lock.

■ 43. To Unload Pistol.—The commands are: 1. UNLOAD, 2. PISTOL. At the command PISTOL, withdraw the magazine. Open the chamber. Glance at the chamber to verify that it is empty. Close the chamber. Take the position of RAISE PISTOL and squeeze the trigger. Then insert an empty magazine.

■ 44. To Inspect Pistol.—The commands are: 1. INSPECTION, 2. PISTOL. (The pistol is inspected mounted only at mounted guard mounting. The magazine is not withdrawn.) At the command PISTOL, take the position of RAISE PISTOL. After the pistol has been inspected, or on command, it is returned.

CHAPTER 3

MARKSMANSHIP, KNOWN-DISTANCE TARGETS, DISMOUNTED

	Paragraphs
SECTION I. Preparatory training	45–51
II. Courses to be fired	52–54
III. Conduct of range practice	55–61
IV. Known-distance targets and ranges; range precautions	62–64
V. Small-bore practice	65–69

SECTION I

PREPARATORY TRAINING

■ 45. INSTRUCTION AND PRACTICE.—*a. Relative value.*—(1) Pistol firing is a purely mechanical operation that any man who is physically and mentally fit to be a soldier can learn to do well if properly instructed. The methods of instruction must be the same as are used in teaching any mechanical operation. The soldier must be taught the various steps in their proper order and must be carefully watched and corrected whenever he makes a mistake.

(2) Good shooting is more the result of careful instruction than of mere practice. Unless properly instructed, men instinctively do the wrong thing in firing the pistol. They instinctively jerk the trigger which is the cause of flinching. Hence, mere practice fixes the instinctive bad habits.

(3) If, however, a man has been first thoroughly instructed in the mechanism of correct shooting and is then carefully and properly coached when he begins firing, correct shooting habits rapidly become fixed.

(4) The ultimate object of the training is to develop the ability to fire one or more accurate shots quickly, but training must begin with carefully coached slow fire to attain accuracy and be followed by practice that will gradually shorten the time without sacrificing the accuracy.

b. Methods of instruction.—(1) Pistol instruction is divided into two phases, preparatory instruction and range firing.

AUTOMATIC PISTOL, CAL. .45, M1911 AND M1911A1   45

In the preparatory instruction the soldier learns practically all the principles of good shooting. In range firing he cultivates the will power to apply these principles when using ball ammunition until proper, fixed habits have been acquired.

(2) The principles of good shooting are simple and easy to learn except the trigger squeeze, which is difficult to apply to a loaded pistol. To this important item most of the instructor's time will be devoted during the period of range practice.

(3) The six distinct steps in the preparatory instruction are—
  (a) Aiming exercises.
  (b) Position exercises.
  (c) Trigger-squeeze exercises.
  (d) Rapid-fire exercises.
  (e) Quick-fire exercises.
  (f) Examination on preparatory work.

(4) The steps are progressive and must always be taught in proper sequence.

(5) Each of the first five steps begins with a talk by the instructor and a demonstration by a squad which the instructor puts through the exercises that are to constitute the day's work. He shows how the corporal organizes the work in the squad so that no men are idle and how the members of each pair coach one another when they are not under instruction by an officer or a noncommissioned officer. He shows exactly how to execute each of the exercises about to be taken up and explains its purpose and application in pistol shooting.

(6) The instructor who gives these very essential talks and demonstrations may be the organization commander, or he may be a specially qualified officer who has been detailed as instructor. But the actual application of the demonstrated exercises to the men of the command must be by the officers and noncommissioned officers of the organization undergoing instruction.

(7) Instruction must be thorough and must be individual. General instruction of groups of men is not enough. The instructors must see that each man understands each and every point and can apply it.

(8) In peacetime training and in war, when time is available for a complete course of instruction and practice, the blank form shown in paragraph 46c (which should be explained in the first talk) must be kept by each squad leader and by each platoon leader independently. This blank form shows at a glance just how much each man knows about each feature of training and permits concentration of instruction where most needed.

(9) Interest and enthusiasm must be sustained and everything possible should be done to stimulate them. If the exercises are carried out in a manner approximately correct and as a routine piece of work, results will be very disappointing.

(10) It is of utmost importance that the trigger squeeze be explained in such a manner as to give the soldier a clear understanding of how it should be executed.

(11) All authorities on shooting agree that the trigger must be squeezed with a steady increase of pressure. If a man knows when his pistol will go off it is because he suddenly gives the trigger all the pressure necessary. Conversely, if the increase of pressure is steady the man cannot know when the piston will be discharged. Hence, he is instructed to *squeeze the trigger in such a way as not to know just when the hammer will fall*. This does not mean that the process is necessarily a slow one and that it will take a comparatively long time to fire a shot. Through training, a man can reduce the time used in pressing the trigger to as brief a period as 1 second and still press it in such a manner that he does not know just what part of the second the discharge will take place. When the soldier has acquired the ability to squeeze the trigger properly, even though it be very slowly, he soon learns to shorten the time without changing the process.

(12) Whenever a man is in a firing position, whether it be a preparatory instruction or during practice firing, he must have a coach beside him to watch him and point out his errors.

(13) None of the preparatory exercises are executed by command or in unison by a group of men. Instruction is individual at all times. The men are placed in pairs and

alternate in coaching each other. This method gives each man the necessary physical rest without halting the progress of his instruction. He is learning while watching another man and attempting to correct his mistakes.

(14) A great deal of preparatory practice is necessary in order to strengthen the muscles of the hand and arm and to fix the habit of correct trigger squeeze. The periods of exercise should not ordinarily be of long duration. Three or four 10-minute periods per day for a month will produce good results on the range. These periods of instruction can often be held during waits when troops are on maneuvers or field exercises. Some kind of a mark can always be found that will serve as an aiming point.

(15) It is a good plan to have full-sized pistol targets placed in the vicinity of the barracks to encourage the men to spend part of their time in preparatory practice.

(16) The preparatory exercises should be held out of doors with full-sized pistol targets, but during inclement weather they can be held indoors, using miniature targets, with good results.

  c. *Scope of preparatory instruction.*—(1) Each man's pistol is closely examined for defects before the beginning of the preparatory instruction.

(2) Every man who is to fire on the range should be put through the preparatory course. Part of the preparatory instruction may have escaped the men the previous year and part of it has certainly been forgotten; in any case it will be beneficial to go over it anew and refresh the mind on the subject.

(3) In peace noncommissioned officers should be put through a rigid test before the period of preparatory instruction for the organization begins. This is also desirable in war when time is available.

■ 46. FIRST STEP; AIMING.—*a. Apparatus required.*—The apparatus required for a set of equipment is listed below. When an entire squad is engaged in this work there should be two sets of this equipment in order that a number of the men do not remain idle. The work of the squad can then be carried on as in rifle marksmanship.

One sighting bar.
One pistol rest.
Two small aiming disks.
One 5-inch aiming disk.
Two small boxes, with paper tacked on one side.
One piece of paper at least 2 feet square and tacked on a wall or frame.

NOTE.—Men who have once been instructed in the aiming exercises, either in preparation for rifle or for pistol firing, will require very little instruction in aiming during subsequent seasons. They will, however, go through the aiming exercises at least once to verify their knowledge of this subject and to assign them a mark in the proper column on the blank form shown in *c* below.

(1) *Sighting bar.*—(a) The sighting bar is illustrated in figure 7.

(b) Carefully blacken all pieces of tin or cardboard and the top of the bar. Nail the bar to a box about 1 foot high and place the box on the ground, table, or other suitable place.

(c) The sighting bar is used in instruction for two reasons, the sights are larger than on the pistol and errors in aiming can be seen more easily and pointed out to the beginner, and the eyepiece of the sighting bar forces the man under instruction to place his eye so that he sees the sights in proper alinement and thus he learns how to aline properly the sights of the pistol. Without an eyepiece the instructor cannot know whether or not the recruit has his eye in proper position.

(2) *Pistol rest.*—(a) To construct a sighting rest for the pistol (fig. 8) use a piece of wood about 10 inches long, $1\frac{1}{4}$ inches wide and $\frac{9}{16}$ inch thick. Shape one end so that it will fit snugly in the handle of the pistol when the magazine has been removed. Screw or nail this stick to the top of a post or other object at such an angle that the pistol when placed on the stick will have its barrel approximately horizontal. A suitable sighting rest for the pistol may be easily improvised by cutting an additional notch to hold the pistol in one end of the box used as a rifle rest.

(b) Having first learned the principles of aiming by means of the sighting bar, the soldier is taught to apply them to the pistol on its rest.

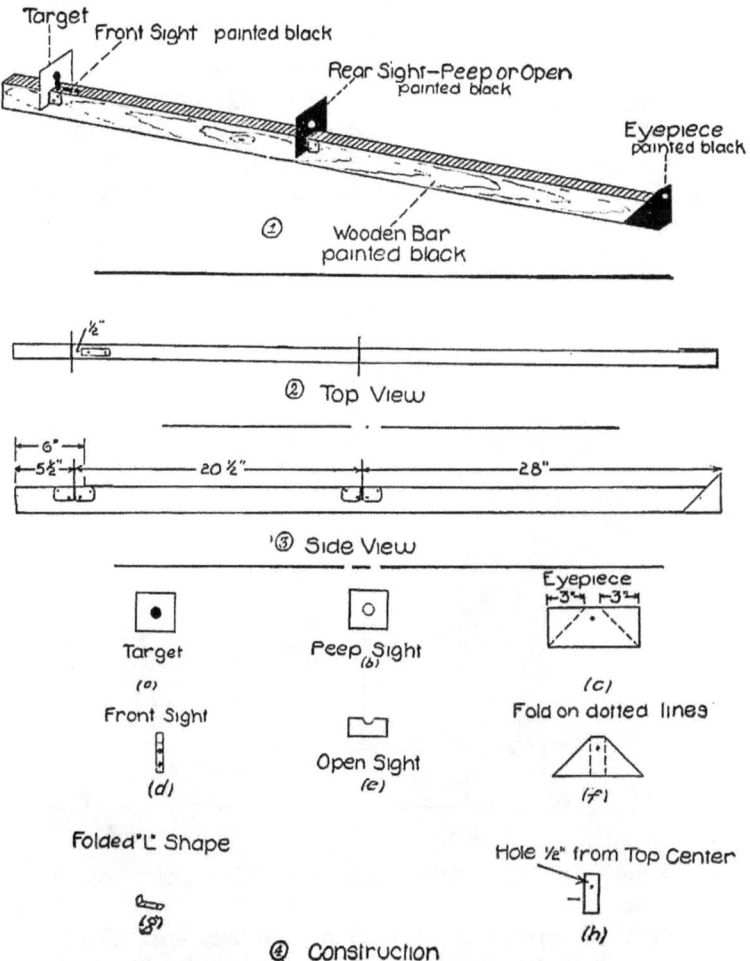

FIGURE 7.—Construction of sighting bar.

Wooden bar—1 by 2 inches by 4 feet 6 inches (approximate).
Eyepiece—Thin metal, 3 by 7 inches; hole, 0.03-inch diameter.
Rear sight—Thin metal or cardboard, 3 by 3 inches; hole in center, ¾-inch diameter.
Front sight—Thin metal, ½ by 3 inches, bent L shape.
Target—Thin metal, or cardboard, 3 by 3 inches, painted white—Black bull's-eye, ¾-inch diameter in center.
Slits—1 inch deep, may be lined with thin metal strips.

(3) *Aiming disks.*—(a) For each sighting bar and each pistol rest a small disk (3 inches in diameter) is made of white cardboard or of tin with white paper pasted on it and with a small bull's-eye in the center. In the exact center of the bull's-eye is a small hole just large enough to admit the point of a pencil. For indoor or close-range work the bull's-eye should not be larger than a 50-cent piece.

(b) There should be one 5-inch aiming disk for each squad for shot-group exercise at 25 yards. The large disk should be of tin, painted black, with a handle 4 or 5 feet long and

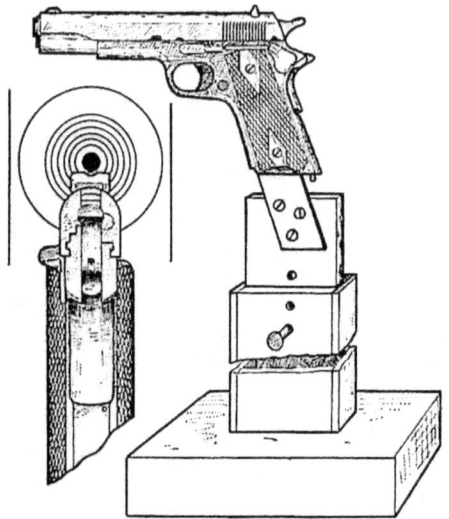

FIGURE 8.—The pistol rest.

of the same color as the paper on which the shot groups are to be made.

b. *Sighting exercises.*—(1) *First exercise.*—(a) The squad leader or instructor shows a sighting bar to his squad or group and points out the front and rear sights, the eyepiece, and the removable target. He explains the use of the sighting bar as follows:

> 1. The front and rear sights on the sighting bar represent enlarged pistol sights.

2. The sighting bar is used in the first sighting exercise because with it small errors can be easily seen and explained to the pupil.
3. The eyepiece requires the pupil to place his eye in such position that he sees the sights in exactly the same alinement as seen by the coach.
4. There is no eyepiece on the pistol, but the pupil learns by use of the sighting bar how to aline the sights properly when using the pistol.
5. The removable target attached to the end of the sighting bar is a simple method of readily alining the sights on a bull's eye.

(b) The instructor explains the open sight to the assembled group, showing each man the illustration of a correct sight alinement (fig. 9).

FIGURE 9.—Normal open sight.

(c) The instructor, using the open sight, adjusts the sights of the sighting bar with target removed to illustrate a correct alinement of the sights. He has each man of the assembled group look through the eyepiece at each of the sight adjustments.

(d) The instructor adjusts the sights of the sighting bar with various small errors in sight alinement and has each man of the assembled group endeavor to detect the errors.

(e) The instructor describes a correct aim, again showing the illustration to each man (fig. 9). He explains that the top of the front sight is seen through the middle of the open sight and is raised to a height so that its top is level with the outside edges of the open sight and just touches the bottom of the bull's-eye so that all of the bull's-eye can be clearly seen.

46   AUTOMATIC PISTOL, CAL. .45, M1911 AND M1911A1

(*f*) The instructor explains that the eye should be focused on the bull's-eye in aiming, and he assures himself by questioning the pupils that each man understands what is meant by focusing the eye on the bull's-eye.

(*g*) The instructor adjusts the sights of the sighting bar and the removable target so as to illustrate a correct aim and has each man of the group look through the eyepiece to observe this correct aim.

(*h*) The instructor adjusts the sights and the removable target of the sighting bar so as to illustrate various small errors and has each man in the group attempt to detect the error.

(*i*) The exercise described above having been completed by the squad leader or other instructor, the men are placed in pairs and the exercise is repeated by the coach-and-pupil method.

(2) *Second exercise.*—(*a*) With the pistol on the pistol rest and the sights pointing at a blank sheet of paper on a board or on the wall, stand with the head in the same relative position as in firing the pistol and look through the sights (fig. 10). Then by signal or by word have the disk moved until the bottom edge of the bull's-eye is in exact alinement with the sights. Then command HOLD and move away from the pistol and let the man undergoing instruction look through the sights to see the proper aim.

(*b*) Have the man under instruction look through the sights while he directs the disk to be moved until the sights are alined on the bottom of the bull's-eye. The instructor then looks through the sights to see if any error has been made.

(*c*) Have the sights adjusted on the bull's-eye with various very slight errors and see if the man under instruction can detect them readily.

(3) *Third exercise.*—Using the sighting rest for the pistol, require the man under instruction to direct the marker to move the disk until the sights are aimed at the bottom edge of the bull's-eye and to command HOLD. The instructor then looks at the aim, and after noticing whether the aim is right or wrong commands: MARK. The marker, without moving the disk, makes a pencil mark on the paper through the hole

in the center of the bull's-eye. Repeat the operation until three marks have been made. The instructor looks at the aim each time, but he says nothing to the man until all three marks have been made and joined together so as to make a shot group. The faults, if any, are then pointed out. The size and shape of the shot group are discussed and the exercise is repeated several times. At 30 feet, using the small bull's-eye, the shot group should be small enough to be covered by a dime.

c. *Blank form.*—This form is used during the period of preparatory instruction. Its object is to show at all times the state of instruction of each man and to insure his thorough instruction in all necessary points before range practice begins.

Names	Recruit instruction				Pistol marksmanship—dismounted										Remarks
	Functioning and operation	Safety precautions	Test of safety devices	Care and cleaning	Sighting bar	Exercises with the pistol rest	Holding the breath	Position of the hand	Position exercise	Trigger squeeze exercise	Instruction in calling the shot	Rapid-fire exercise	Quick-fire exercise	Ability as a coach	Final examination	
------	----	----	----	----	----	----	----	----	----	----	----	----	----	----	----	
------	----	----	----	----	----	----	----	----	----	----	----	----	----	----	----	
------	----	----	----	----	----	----	----	----	----	----	----	----	----	----	----	

Method of marking: [×] Fair.   [× ×] Good.

[× × ×] Very good.   [× × × ×] Excellent.

[× × × × ×] Excellent. Has instructional ability.

■ 47. SECOND STEP; POSITION.—*a. Essentials of proper position.*—To assume the proper position for firing it is necessary to know how to aim, how to grasp the pistol, how to hold the breath properly, and the correct position of the body with relation to the target.

(1) *How to grasp the pistol.*—(a) To take the grip, hold the pistol in the left hand and force the grip safety device down and back into the crotch formed between the thumb and forefinger of the right hand. The thumb is carried parallel with or slightly higher than the forefinger; it should

FIGURE 10.—Sighting exercise.

never be lower. Close the three lower fingers on the stock firmly but not with a tense grip (fig. 11).

(b) The thumb and forefinger squeeze the frame of the pistol, but the ball of the thumb does not always touch the pistol, depending on the conformation of the man's hand. By this pressure movement to the right or left is controlled, and the trigger squeeze can be better applied and coordinated.

AUTOMATIC PISTOL, CAL. .45, M1911 AND M1911A1   47

(c) The muscles of the arm are firm without being rigid. There should be no bending of the arm at the elbow when the pistol is fired. On the other hand, the arm should not be locked at the elbow. When the firer is shooting properly, after recoil the pistol arm should automatically carry the pistol back to the position shown in figure 12.

(2) *How to hold the breath.*—(a) The proper method of holding the breath is important because without instruction many men hold the breath in the wrong way or do not hold it at all.

(b) To hold the breath, draw into the lungs a little more air than an ordinary breath, let a little of the air out and stop the rest by closing the throat. Do not hold the breath

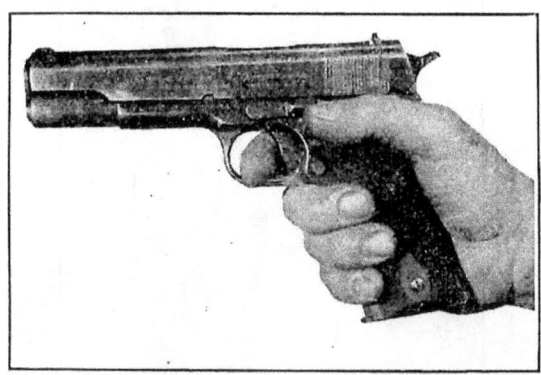

FIGURE 11.—How to grasp the pistol.

with the throat open or by the muscular effort of the diaphragm.

(3) *Position of the body* (fig. 12).—(a) The position of the body is a little more than half faced to the left, the feet 12 or 18 inches apart, depending on the man, the head erect, and the body perfectly balanced when the pistol is held in the shooting position.

(b) The whole position should be natural and comfortable. Upon assuming the position there is some point to which the pistol points naturally and without effort. If this point is not the center of the target the whole body must be shifted so as to bring the target into proper alinement. Otherwise the firer will be firing under a strain because he will be pulling the pistol on the target by muscular effort for each shot. Any unnecessary tensing of any of the muscles of the hand, arm, or body will cause tremors and should therefore be avoided.

b. Position exercise.—(1) Required for this exercise: A line of L targets with firing points at 15 and 25 yards, or a line of small aiming bull's-eyes placed at the height of the shoulder.

(2) The men, armed with the pistol, are placed in one line at 1-pace intervals. Give the command: 1. INSPECTION, 2. PISTOL, and verify the fact that all pistols are unloaded.

FIGURE 12.—Position of the body.

(3) Demonstrate the position of the hand in gripping the stock and describe the grip in detail.

(4) Require each man to grip the stock of his pistol in the prescribed manner, using the free hand to grasp the barrel and set the stock well back in the pistol hand between the thumb and the first finger.

(5) Describe the correct method of holding the breath while aiming and require each man to practice it a few times.

(6) Demonstrate the correct position of the whole body when firing, explaining in detail the position of the feet, legs, body muscles, arms, and head.

(7) Require each man to assume the correct firing position. The officers and noncommissioned officers of the organization correct individuals who are at fault.

(8) The above exercises having been completed, instruction becomes individual under a coach. The men are placed in pairs opposite L targets or opposite small aiming bull's-eyes and take turns coaching each other.

(9) The details of the position exercises are—
(a) Grasp the stock with the correct grip.
(b) Face target, then face half left.
(c) Separate the feet 12 to 18 inches.
(d) Aline the sights on the bottom edge of the bull's-eye, arm extended.
(e) Hold the breath.
(f) As soon as the arm becomes tired or the aim becomes unsteady, assume the position of RAISE PISTOL.
(g) The pistol should be removed from the right hand and the muscles of the hand, arm, and shoulder relaxed and exercised before resuming the grip. This should also be done between shots in slow fire.

(10) After the firer has completed the position exercise he may repeat it with a weight, such as a pair of field glasses in a case suspended from the right arm. The weight is suspended first between shoulder and elbow, then from the forearm, then from the wrist, and finally from the barrel of the pistol, interspersed with short rests. The value of this exercise lies in developing the muscles of the shoulder and arm.

(11) (a) The hammer is not raised during the position exercises and the trigger is grasped very lightly with the finger.
(b) After a short rest repeat the exercise.
(c) The man acting as coach watches carefully and corrects all errors.
(d) The man under instruction and the coach change places as the officer in charge of the instruction desires. This should be every 3 to 5 minutes.

(e) Only a few hours in all should be devoted to the position exercises, as all of its details are included in the trigger-squeeze exercise.

■ 48. THIRD STEP; TRIGGER SQUEEZE.—*a. Importance of correct trigger squeeze.*—(1) The recruit can readily learn to aim and hold the aim either on the bull's-eye or very close to it for at least 10 seconds. When he has learned to press the trigger in such a manner as not to spoil his hold he becomes a good shot. All men flinch in firing the pistol if they know the exact instant at which the discharge is to take place. This is an involuntary action which cannot be controlled. A sudden pressure of the trigger may derange the aim slightly but the extreme inaccuracy of a shot fired in this way is due mainly to the flinch; that is, the thrusting forward of the hand to meet the shock of recoil. Any man who holds the sights of the pistol as nearly on the bull's-eye as possible and continues to press on the trigger with a uniformly increasing pressure until the pistol goes off is *a good shot*. Any man who has learned to increase the pressure on the trigger only when the sights are in alinement with the bull's-eye, who holds the pressure when the muzzle swerves, and who continues with the pressure when the sights are again in line with the bull's-eye is an *excellent shot*. Any man who tries to "catch his sights" as they touch the bull's-eye and to set the pistol off at that instant is a *very bad shot*.

(2) The apparent unsteadiness of the pistol while being held on the bull's-eye does not cause much variation in the striking place of the bullet due to the fact that the movement is of the whole extended arm and pistol. But the sudden pressure of the trigger which always accompanies the flinch deflects the muzzle of the pistol and causes the bullet to strike far from the mark. In squeezing the trigger the pressure must be *straight* to the rear. There is a tendency on the part of some men to press the trigger also to the left.

b. Calling the shot.—To call the shot is to state where the sights were pointed at the instant the hammer fell; thus, "high," "a little low," "to the left," "slightly to the right," "bull's-eye," etc. If the soldier cannot call his shot correctly in range practice he did not press the trigger properly and consequently did not know where the sights were pointed when the hammer fell.

c. Trigger-squeeze exercises.—(1) *First exercise.*—(a) Required for this exercise: A line of L targets with a firing point at 25 yards.

(b) Give the command: 1. INSPECTION. 2. PISTOL, and verify the fact that all pistols are unloaded.

(c) The squad leader explains to his squad the details of this exercise which are—
    1. Cock and lock the piece.
    2. Take the correct grip.
    3. Take the correct position.
    4. Aline the sights on the target and start the squeeze, gradually increasing the pressure on the trigger until all the strength of the hand is employed.
    5. Rest the hand.
    6. Repeat the above operation with the piece unlocked.
    7. Call the shot.

(d) The squad leader assures himself that all the men understand the details of this exercise. The work is then carried on by pairs working together, coach and pupil. Members of the squad should change over frequently to avoid tiring the muscles of the arm. Extended trigger-squeeze exercise is necessary and the periods should be short but frequent.

(e) The duties of the coach are to—
    1. See that the firer takes the correct grip.
    2. See that the firer takes a correct position.
    3. Watch the hand of the firer to see that he is gradually increasing the pressure on the trigger.
    4. See that the firer rests his shooting hand after having unlocked the piece.
    5. See that the firer calls the shot when the exercise is repeated with the piece unlocked.

(2) *Second exercise.*—(a) Give the command: 1. INSPECTION, 2. PISTOL, and verify the fact that all pistols are unloaded.

(b) The squad leader explains to his squad the details of this exercise, which are—
    1. Cock the piece.

2. Take the correct grip.
3. Take the correct position.
4. Aline the sights on the target and start the squeeze. Close the eyes and continue to squeeze until the hammer falls.
5. When the hammer falls, open the eyes and check the aim to see if it has been deranged.

NOTE.—The firer should be able to keep on the target. If he is off persistently he should check on his grip and position to see that they are correct.

(c) The duties of the coach in this exercise are the same as in the first trigger-squeeze exercise.

(3) *Third exercise.*—In learning to fire the pistol, the average man has a tendency to push or punch forward with the arm or shoulder to meet the force of recoil of the piece. To assist the firer in overcoming this tendency the following exercise is prescribed: By command, verify the fact that all pistols are unloaded. The firer cocks and locks the piece, takes his firing position, alines the sights under the bull's-eye, and squeezes the trigger. The coach stands in front of the firer, facing him, and strikes the muzzle of the piece with the palm of his hand. At irregular intervals his hand misses the muzzle. The firer should hold the piece on the target and make no forward punching movement to meet the shock of the blow. If he does push forward with the arm or shoulder it will be apparent when the coach misses the muzzle.

■ 49. FOURTH STEP; RAPID FIRE.—*a. Training for rapid fire.*—(1) Training for rapid fire is taken up after the trigger-squeeze exercise has been practiced sufficiently to be understood thoroughly, but the trigger-squeeze exercise practice in slow fire should be resumed and continued during the entire period of preparatory training.

(2) The time consumed in pressing the trigger must necessarily be shorter in rapid fire than in slow fire, but the process is the same.

(3) To fire the first shot, the pistol should be brought from the position of RAISE PISTOL by the shortest route to the aiming position with the sights alined on the mark. This is done by a smooth, rapid extension of the right arm straight from the

shoulder, inserting the right forefinger in the trigger guard during the movement and holding the breath. To bring the pistol through the arc of a circle to the aiming position is an unnecessary loss of valuable time.

(4) For succeeding shots the sights should be held as nearly on the mark as possible and the breath held throughout the score. The recoil after each shot will throw the sights out of alinement, but they should be brought back immediately to the mark by the shortest route. The recoil should cause a vertical movement of the firing arm upward, the hand moving only 6 to 8 inches. There should be no snapping or bending of the wrist or elbow. The sights will then come back automatically on the mark after each shot. To give the pistol a flourish between shots is a useless loss of time.

(5) To simulate the self-loading action of the automatic pistol take a strong cord about 4 feet long and tie one end to the thumbpiece of the hammer, *the knot on top.* Take a few turns of the other end of the cord around the thumb or fingers of the left hand. The cord should be long enough to permit the left hand to hang naturally at the side while aiming the pistol with the right hand, right arm fully extended.

(6) Each time the hammer falls a quick backward jerk of the left hand recocks the pistol and at the same time jerks the sights out of alinement with the bull's-eye. This derangement of the alinement corresponds very closely to the jump of the pistol when actually firing.

(7) If the knot is underneath the hammer or if a very thick cord is used the hammer will not remain cocked when jerked back.

b. Rapid-fire exercise.—(1) Required for this exercise: A piece of strong cord 4½ feet long for each man; a row of L targets or a row of aiming bull's-eyes.

(2) Give the command: 1. INSPECTION, 2. PISTOL, and verify the fact that all pistols are unloaded.

(3) Explain to the assembled command that the trigger squeeze is the same in rapid fire as in slow fire.

(4) Demonstrate the correct method of bringing the pistol by the shortest route to the aiming position. Show how this is done from RAISE PISTOL and in drawing the pistol from the holster in an emergency.

AUTOMATIC PISTOL, CAL. .45, M1911 AND M1911A1

(5) Show how to tie the cord to the thumbpiece of the hammer and cause each man to adjust his cord.

(6) Demonstrate the method of cocking the pistol by means of the cord and explain how this simulates the self-loading action of the pistol.

(7) Show how the pistol is kept as nearly on the mark as possible during the whole score. Caution the men to avoid unnecessary flourishes or movements between shots.

(8) Demonstrate—

(a) The action of the pistol in recoil when a shot is fired.

(b) How the arm should not be permitted to bend at the elbow.

(c) How the pistol should move upward through a small arc and be deflected from the original point of aim only a short distance.

(d) How the forefinger should move forward after the explosion only far enough to allow the sear to become re-engaged and immediately thereafter start pressing the trigger for the next shot.

(e) How the eye should not be allowed to close when the explosion occurs.

(f) How the breath should be held for each shot.

(9) (a) The above demonstrations having been completed, the men are placed in front of the line of targets in pairs, one to practice and one to coach. The exercise is then carried on exactly the same as rapid fire in range practice. If a line of disappearing targets has been arranged for this exercise the targets appear, remain in sight the allotted time, and then disappear. While the targets are in sight each man undergoing instruction attempts to fire five shots (simulated fire), cocking the piece for each shot except the first by a jerk of the cord with the left hand.

(b) If the targets are stationary the exercise begins with the command: 1. COMMENCE, 2. FIRING, and ends with the command: 1. CEASE, 2. FIRING.

(c) After each three or four scores of simulated fire the men of each pair are directed to change places, the firer becoming the coach and the coach becoming the firer.

(10) (a) In this exercise the coach carefully watches the man and corrects all errors in grip, position, trigger squeeze,

and manipulation of the piece, paying particular attention to the trigger squeeze.

(b) Rapid-fire exercises should be frequent but not of long duration.

(c) It is advisable to extend the time limits several seconds when rapid-fire exercise is first taken up. The time limit is then gradually reduced until it corresponds to the time prescribed for range firing, record practice.

■ 50. FIFTH STEP; QUICK FIRE.—a. *Training for quick fire.*—(1) The training for quick fire is taken up after the rapid-fire exercise has been practiced sufficiently to be understood thoroughly. Thereafter, exercises in slow fire, rapid fire, and quick fire should all be continued until the end of the period of preparatory training.

(2) For each shot the pistol is brought from RAISE PISTOL to the aiming position by the shortest route after the target appears.

(3) The pistol may be cocked after each shot in this exercise either by means of a cord as in rapid-fire exercise, or by using the left hand to pull the hammer back after the position of RAISE PISTOL is resumed.

b. *Quick-fire exercise.*—(1) Required for this exercise: A line of E targets that can be operated as bobbing targets from a pit or screen, or a line of E targets so arranged on pivots that the edge can be turned toward the firer when the target is not exposed.

(2) Give the command: 1. INSPECTION, 2. PISTOL, and verify the fact that all pistols are unloaded.

(3) Explain to the assembled command that the trigger squeeze is the same in quick fire as in slow fire.

(4) Demonstrate the correct method of bringing the piece from RANGE PISTOL to the aiming position.

(5) Show how the pistol is cocked between shots when the left hand is used instead of the cord.

(6) The above demonstrations having been completed, the men are placed in pairs in front of the line of bobbing targets, one man of each pair to act as coach for the other man. The exercise is then carried on exactly the same as quick fire in range practice. The targets appear, remain in sight the allotted time, and then disappear. After the targets appear

each man undergoing instruction brings his pistol from RAISE PISTOL to the aiming position, aims, fires one shot (simulated fire), and returns his piece to the position of RAISE PISTOL. After three or four scores of simulated fire the men of each pair are directed to change places.

(7) The coach watches carefully the man going through the exercises and corrects all errors in the grip, position, holding the breath, trigger squeeze, and the manipulation of the piece, paying particular attention to the trigger squeeze. It is advisable to extend the time limit about two seconds for each shot when quick-firing exercise is first taken up. The time is then gradually reduced until it corresponds to the time prescribed for range firing, record practice.

(8) When disappearing targets cannot be provided for this exercise it may be held with stationary E targets. The command UP is given to signify that the targets are in sight, and the command DOWN to signify that they have been withdrawn.

(9) Practice in quick fire should be held frequently, but the periods of practice should not be of long duration.

(10) If the range is some distance from the area designated for preparatory exercises, or it is impracticable to arrange for a line of bobbing targets, L targets may be substituted for the bobbing targets.

■ 51. SIXTH STEP; EXAMINATION.—At the completion of the preparatory instruction, the instructor should assure himself by an examination that every man understands thoroughly and can explain every phase of the preparatory training. The questions and answers given below are merely examples. Each man should be required to explain each item in his own words.

Instructor: Examine your pistol to see that it is unloaded.

Q. What are the safety devices of the pistol? A. The safety lock, the grip safety, the half-cock notch, and the disconnector.

Q. Show me how you test the safety lock. A. I cock the pistol, move the safety lock up into place, and then grip the stock and see if the hammer remains up when pressure is applied to the trigger.

Q. Show me how you test the grip safety. A. I cock the pistol, see that the safety lock is down and then, without putting any pressure on the grip safety, I see if the hammer

AUTOMATIC PISTOL, CAL. .45, M1911 AND M1911A1   51

will remain up when a strong pressure is applied to the trigger.

Q. Show me how you test the half-cock safety device. A. I half cock the pistol, grip the stock, and see if the hammer remains at half cock when pressure is applied to the trigger. Then I take my finger off the trigger, pull the hammer back almost to full cock, and let go of it to see if it stops at half cock as it falls.

Q. Show me how you test the disconnector. A. I cock the pistol and grip the stock; then with my left hand I move the slide to the rear a quarter of an inch; I then apply a strong pressure on the trigger and release the slide to see if the hammer will remain up. I also pull the slide fully back until it is held in place by the slide stop; I then grip the stock, apply a strong pressure on the trigger and release the slide by pressing down the slide stop with my left hand. The hammer should remain up after the slide moves forward into place.

Q. If the hammer does not remain up after the slide moves forward into place, what does it indicate? A. That with ball ammunition the pistol would continue to fire automatically as long as pressure is maintained on the trigger, which is very dangerous.

Q. If any of the tests of the safety devices fail at any time, what should you do? A. I should report the matter at once to my platoon or company commander.

Q. What is this (indicating a sighting bar)? A. A sighting bar.

Q. What is it used for? A. To teach men how to aim.

Q. Why is it better than a pistol for this purpose? A. Because the sights are much larger and slight errors can be seen more easily and pointed out.

Q. What does this represent? A. The front sight.

Q. What does this represent? A. The rear sight.

Q. What is this? A. The eyepiece.

Q. What is it for? A. To make the man hold his head in the right place so that he will see the sights properly alined.

Q. Is there an eyepiece on the pistol? A. No. A man learns by the sighting bar how the sights look when properly alined, and he must hold the pistol while aiming so as to see the sights in the same way.

Q. Adjust the sights of this sighting bar so that they are in proper alinement with each other. (Verified by instructor.)

Q. Now that the sights are properly adjusted, have the small bull's-eye moved until the sights are aimed at it properly. (Verified by instructor.)

Q. Tell me what is wrong with this aim. (The instructor now adjusts the sights of the sighting bar on the bull's-eye with various very slight errors, requiring the man to point out the error.)

Q. Show me how you grip the stock of the pistol.

Q. Show me the position you take when you are going to shoot.

Q. How do you squeeze the trigger? A. I squeeze it with such a steady increase of pressure as not to know exactly when the hammer will fall.

Q. If the sights get slightly out of alinement while you are squeezing the trigger, what do you do? A. I hold the pressure I have on the trigger and only go on with the increase of pressure when the sights become alined again.

Q. If you do this can your shot be a bad one? A. No.

Q. Why? A. Because I cannot flinch for I do not know when to flinch, and the sights will always be lined up with the bull's-eye when the shot is fired because I never increase the pressure on the trigger except when the sights are properly alined.

Q. When you are practicing in slow fire and your arm becomes unsteady and your aim uncertain, what should you do? A. I should come back to RAISE PISTOL without firing the shot and then try again after a short rest.

Q. If it is impossible for you to hold the pistol very steady, can you still do good shooting? A. Yes; if I press the trigger properly.

Q. Tell me why that is. A. Because the natural unsteadiness of the arm moves the whole pistol and the barrel remains nearly parallel to the line of sight. But if I give the trigger a sudden pressure the front end of the barrel will be thrown out of line with the target, and the bullets will strike far out from the mark.

AUTOMATIC PISTOL, CAL. .45, M1911 AND M1911A1   51

Q. What causes this deflection of one end of the pistol when the trigger is given a sudden pressure? A. The sudden pressure itself causes some of it, but most of it is caused by the flinch that always accompanies this kind of a trigger pressure.

Q. What does a man do when he flinches in shooting a pistol? A. He usually thrusts his hand forward is if trying to meet the shock by suddenly stiffening all his muscles.

Q. Must the trigger always be squeezed slowly in order to do it correctly? A. No. I squeeze it the same way in rapid fire and quick fire. The time is shorter but the process is the same.

Q. What is meant by calling the shot? A. To say where you think the bullet will hit as soon as you shoot and before the shot is marked.

Q. How can you do this? A. By noticing exactly where the sights point at the time the pistol is fired.

Q. If a man cannot call his shot correctly, what does it indicate? A. That he did not squeeze the trigger properly and consequently did not know where the sights were pointed at the instant the discharge took place.

Q. Show me how you hold your breath while aiming.

Q. Take your pistol. Aim at that bull's-eye and squeeze the trigger a few times, calling the shot each time. (The instructor particularly notes the holding of the breath.)

Q. Show me how you come to a position of aim from RAISE PISTOL.

Q. Show me how you come to the aiming position in drawing the pistol from the holster in an emergency.

Q. Take this pistol with the cord tied to the hammer and fire a rapid-fire score at that target (simulated fire).

Q. Fire a score (simulated fire) at that quick-fire target. I will give the command UP when it is supposed to come into sight, and the command DOWN when it is supposed to be withdrawn from view.

Q. What do you do in case a cartridge misses fire? A. I bring the piece to RAISE PISTOL, grasp the slide with my left thumb and finger, pull the slide fully back and let go of it.

This throws out the faulty cartridge and loads in another cartridge.

Q. Are there any points about pistol firing that you do not understand?

NOTE.—In all the demonstrations by the man undergoing examination the instructor carefully notes all points that are covered in the preparatory exercises. Each man is put through a thorough test along the line indicated in these questions and answers before he is allowed to fire.

SECTION II

COURSES TO BE FIRED

■ 52. GENERAL.—AR 775-10 prescribes details as to who will fire and ammunition allowances.

■ 53. INSTRUCTION PRACTICE.—The following tables prescribe the firing in instruction practice in the order followed by the individual soldier. Target L is used in much of the practice as the bull's-eye makes competition keener and shows up errors as no other target can.

a. Slow fire.

TABLE I.—*Slow fire—Target L*

Range	Time	Scores (5 shots each), minimum
15 yards	No time limit	1
25 yards	----do	1

Unlimited time is permitted for slow fire in order to permit proper explanation of the causes of errors and indication of corresponding remedies. It is intended to be the elementary phase of instruction in the proper manipulation of the weapon and for determining and correcting the personal errors of the firer.

AUTOMATIC PISTOL, CAL. .45, M1911 AND M1911A1   53–54

b. *Rapid fire.*

TABLE II.—*Rapid fire—Target L*

Range	Time	Scores (5 shots each), minimum
15 yards	11 seconds per score	1
25 yards	15 seconds per score	1

If pits are used, time is taken at the pits as in rapid fire rifle practice. If pits are not used time is taken at the firing point. The target being up, the soldier stands with the weapon at RAISE PISTOL, loaded and locked. The command: 1. COMMENCE, 2. FIRING, is given and the soldier must fire one score within the prescribed limit of time, at the end of which the command: 1. CEASE, 2. FIRING, will be given. Intervals of time are measured from the last words of the commands.

c. *Quick fire.*

TABLE III.—*Quick fire—Target E—Bobbing*

Range	Time	Scores (5 shots each), minimum
15 yards	2 seconds per shot	2
25 yards	3 seconds per shot	2

■ 54. RECORD PRACTICE.—The following tables prescribe the firing in record practice in the order followed by the individual soldier. The procedure is as in instruction practice.

a. *Slow fire.*

TABLE IV.—*Slow fire—Target L*

Range	Time	Scores (5 shots each)
25 yards	No time limit	2

51

b. *Rapid fire.*

TABLE V.—*Rapid fire—Target L*

Range	Time	Scores (5 shots each)
15 yards	11 seconds per score	2
25 yards	15 seconds per score	2

c. *Quick fire.*

TABLE VI.—*Quick fire—Target E—Bobbing*

Range	Time	Scores (5 shots each)
25 yards	3 seconds per shot	3

SECTION III

CONDUCT OF RANGE PRACTICE

■ 55. COACHING METHODS.—*a. Range practice.*—(1) The object of range practice is to teach the men to apply with a loaded pistol the principles of good shooting that they have learned during the preparatory exercises.

(2) Each man while firing must have a coach to correct him whenever he violates any of these principles.

(3) Slow-fire practice should be carried on until the man under instruction thoroughly understands the principles of good shooting.

(4) When rapid fire and quick fire are first taken up the time limit should be extended a few seconds. The time should then be gradually reduced until the scores are being fired in the time prescribed for record practice.

b. Dummy cartridges.—(1) Dummy cartridges are of great value in teaching both slow and rapid fire.

AUTOMATIC PISTOL, CAL. .45, M1911 AND M1911A1   55

(2) *Dummy cartridges must not be used except on the firing line of the pistol range.* The same precautions are observed as in using service ammunition.

c. *Slow fire.*—(1) The coach stands on the left side of the firer in such a position as to be able to observe the latter's trigger finger, his grip, his eye, and his position. It is the duty of the coach to correct all errors. The coach fills the magazines for the firer and hands them to him. At the beginning of range practice the magazines should be filled partly with service ammunition and partly with dummy cartridges. The firer must not know how many dummy cartridges are in the magazine or the order in which they are packed.

(2) The object of placing dummy cartridges in the magazine is to show the coach whether or not the man under instruction is squeezing the trigger correctly, and in case of an improper trigger squeeze to bring the fact forcibly to the attention of the firer himself. When a loaded cartridge is fired the flinch is often masked by the recoil of the pistol and the firer is not conscious of having flinched. When the hammer falls on a dummy cartridge which the firer thinks is loaded, the sudden stiffening of the muscles and the thrusting forward of the hand to meet the shock that does not come are apparent to everybody in the vicinity, including the firer himself. The mixing of dummy cartridges with service ammunition causes the man to make a determined effort to press the trigger properly for all shots.

(3) The firing of scores with dummy cartridges and service ammunition should not be confined to the early stages of training. It is advisable to have some practice of this kind each day during the entire period of instruction practice. Many expert pistol shots use this form of practice in training for competitions.

(4) The following items of instruction are given to a pupil on beginning range practice, even though that person has already done a great deal of shooting. Once a person has been put through this instruction it is usually not necessary to repeat it during subsequent periods of range practice.

(a) Explain the method of grasping the piece.

53

(b) Show the amount of force used in gripping the stock by grasping the pupil's hand, saying: "This is too tight a grip" (gripping his hand very tightly). "This is too loose a grip" (gripping his hand loosely). "This is the right amount of force to use in gripping the stock" (gripping his hand with the firm but comfortable grip that should be used in shooting).

(c) Explain and demonstrate the position of the body, the feet, and the arm (par. 47 and fig. 12), and have the pupil assume this position.

(d) Explain the proper method of aiming.

(e) Explain that any man can aim and hold well enough for a good score. Have the pupil assume the proper position and aim at the target with an empty pistol without attempting to press the trigger to see how long he can hold the sights on or near the bull's-eye. Explain to him that this aiming at the target with an empty gun demonstrates how near to the center his bullets will strike provided he presses the trigger properly.

(f) Explain the proper method of pressing the trigger.

(g) Have the pupil aim at the target with an empty pistol and then press the trigger for him several times as described in *d* below (fig. 13), directing the pupil to call the shot each time the hammer falls.

(h) Have the pupil aim at the target with a loaded pistol and then press the trigger for him as described in *d* below, directing him to call the shot each time the piece is fired. Fire a score of five shots in this way.

(i) Have the pupil fire a score of five shots, pressing the trigger himself to see if he can press the trigger properly and make as good a score as the one made when the coach pressed the trigger.

d. Squeezing the trigger.—One method of showing the men under instruction how to squeeze the trigger properly is to have him hold and aim the pistol while the coach presses the trigger. This is done in the following manner:

(1) The coach demonstrates the value of correct trigger press to the student by placing his hands in the position shown in figure 13 and pressing on the end of the pupil's trigger finger with his left thumb. The coach cautions the

pupil neither to assist nor resist the pressure which is put on the end of his trigger finger but to devote his whole attention to his aim and hold.

(2) The coach must be careful to apply a slow, steady pressure to the finger of the pupil and at the same time not interfere with the pupil's aim while applying this pressure. As a rule, the coach should consume from 5 to 10 seconds in putting sufficient pressure on the pupil's finger to fire the pistol.

(3) When pressing the trigger for a pupil as above described the coach should hold his head well to the rear to

FIGURE 13.—Coach pressing the trigger.

keep from having his left ear too near the muzzle of the piece.

(4) If the firer shows a tendency to apply the last part of the squeeze himself by giving the trigger a sudden pressure, he is directed to place his finger below the trigger guard, and the coach applies the pressure directly to the trigger instead of through the finger of the man under instruction.

e. *Calling the shot.*—Men should be required to call each shot in slow fire. If a man does not call the shot immediately after firing, the coach directs him to do so.

f. *Coaching rapid fire.*—(1) The firing of scores with dummy cartridges and service ammunition mixed is a very valuable form of rapid-fire practice. The coach fills the magazine in such a way that the firer cannot know the order in which the cartridges are placed.

(2) The coach must watch the man closely, and each time he is seen to flinch, whether on a loaded or a dummy cartridge, the coach should caution him.

(3) When the hammer falls on a dummy cartridge the firer grasps the slide with his left hand, pulls it fully back, and releases it. This ejects the dummy and loads another cartridge. The time limit must be extended to compensate for the time lost in ejecting the dummy cartridges. It should not take more than 2 seconds to eject a dummy cartridge and resume the aiming position.

g. Coaching quick fire.—(1) The use of dummy cartridges and the coaching methods are the same in quick fire as in rapid fire.

(2) The occasional use of dummy cartridges in both rapid fire and quick fire should be continued throughout the entire period of instruction practice.

■ 56. SAFETY PRECAUTIONS ON THE RANGE.—*a.* Never place a loaded magazine in the pistol until you have taken your place at the firing point.

b. Always remove the magazine and unload the pistol before leaving the firing point.

c. Always hold the loaded pistol at the position of RAISE PISTOL, except while aiming.

d. When firing ceases temporarily, lock the piece and hold it at RAISE PISTOL. Do not assume any position except RAISE PISTOL without first removing the magazine and unloading.

e. If one or more cartridges remain unfired at the end of a rapid-fire or quick-fire score, remove the magazine and unload immediately.

■ 57. RANGE ORGANIZATION.—*a.* The work on the range should be so organized that no men are idle for any length of time. A good arrangement is four or six orders per target. It should never be necessary to assign more than six orders. If there is not a sufficient number of targets to provide for this, the extra men should remain off the range and be given other instruction.

b. One method is to have a line of pistol targets on a flank of each firing point of the rifle range, so arranged that the firing points of the rifle range and of the pistol range are on one line. There should be about 50 yards interval between the rifle range and the pistol range. The targets may be placed on the ground instead of in pits. The bobbing targets are arranged to revolve on their own axis and are operated from behind the firing line by means of cords. When the targets are to be marked the whole line ceases firing, unloads pistols, and moves up to the targets to record the hits and paste the shot holes. In slow fire the coach can keep the firer informed as to the location of his hits by use of field glasses.

c. When the time is short and range facilities and proper supervision permit, rifle firing and dismounted pistol firing may be carried on at the same time. While one group is firing on the rifle range the other is firing on the pistol range. As the men complete a score with the rifle they move to the pistol range and their places at the rifle firing point are filled by men who have completed a score of pistol firing. As soon as all men present have completed their scores with the rifle the whole group moves back to the next firing point (moving the pistol targets if necessary) and continue as before with the alternate rifle and pistol firing.

d. The pistol targets may be placed so that the line of fire is at right angles to the line of fire of the rifle range if the terrain permits. When it is not practicable to have pistol firing and rifle firing at the same time, other means will be adopted to keep the men occupied while they are not actually firing or coaching.

■ 58. TARGET DETAILS.—The personnel for the supervision, operation, and scoring of targets during recorded firing consists of officers, noncommissioned officers, and privates as follows:

a. One commissioned officer assigned to each four targets. The officer will take up and sign each duplicate score card as soon as a complete score is recorded.

b. One noncommissioned officer assigned to each target to enter scores on the duplicate score cards and to direct and supervise the detail pasting the target. This noncommissioned officer will be selected, except at a 1-company post, from an organization other than the one firing on the target which he supervises. When the post is garrisoned by a single company so that it is impossible to detail noncommissioned officers of other companies to supervise the marking and scoring, those duties are performed by the noncommissioned officers of the firing company.

c. One or two privates to operate, mark, and paste each target.

d. When the targets are not placed in pits, target details may be reduced to one commissioned officer for each four targets, one noncommissioned officer for not to exceed each two targets, and one private to each target.

e. The noncommissioned officer examines the target after each score is fired and enters the score on the score card, initialing same. He directs the private to paste the target after the score is recorded and marked and examines the target to see that no shot holes are left unpasted.

■ 59. REGULATIONS GOVERNING RECORD PRACTICE.—*a. Coaching prohibited.*—Coaching of any nature, after the firer takes his place on the firing point, is prohibited. No person may render or attempt to render the firer any assistance whatever while he is taking his position or after he has taken his position at the firing point. Each firer must observe the location of his own hits.

b. Shelter for firer.—Sheds or shelter for the firer are not permitted on any range.

c. Cleaning.—Cleaning is permitted only between scores.

d. Gloves.—A glove may be worn on either or both hands.

e. Pieces loaded on command.—Pieces are not loaded except by command or until position for firing has been taken.

f. Shots cutting edge of bull's-eye or line.—Any shot cutting the edge of the figure or bull's-eye is signaled and recorded as a hit in the figure or the bull's-eye. Because the limiting line of each division of the target is the outer edge of the line separating it from the exterior division, a shot touching this line is signaled and recorded as a hit in the higher division.

g. Slow fire score interrupted.—If a slow fire score is interrupted through no fault of the person firing, the unfired shots necessary to complete the score are fired at the first opportunity thereafter.

h. Misses.—In all firing before any miss is recorded the target is carefully examined by an officer.

i. Accidental discharges.—All shots fired by the soldier after he has taken his place on the firing line (and it is his turn to fire, the target being ready) are considered in his score even if his piece was not directed toward the target or is accidentally discharged.

j. Firing on wrong target.—Shots fired upon the wrong target are entered as misses upon the score of the man firing no matter what the value of the hits upon the wrong target may be. In rapid fire the soldier at fault is credited with only such hits as he may have made on his own target.

k. Two shots on same target.—In slow fire, if two shots strike a target at the same time or nearly the same time, and if one of these shots was fired from the firing point assigned to that target, the hit having the highest of the two values is entered on the soldier's score and no record is made of the other hit.

l. Withdrawing target prematurely.—In slow fire, if the target is withdrawn from the firing position just as the shot is fired, the scorer at that firing point at once reports the fact to the officer in charge of the scoring on that target. That officer investigates to see if the case is as represented. Being satisfied that such is the case, he directs that the shot be not considered and that the man fire another shot.

m. Misfires.—In case of a misfire in rapid fire or quick fire the soldier ceases firing and takes the position of RAISED PISTOL. In rapid fire the target is not marked and the score is repeated. In quick fire the score is continued after the defective cartridge has been replaced.

n. Unused cartridges in rapid fire.—Each unfired cartridge is recorded as a miss.

o. Disabled pistol.—If during the firing of a rapid or quick fire score, the pistol becomes disabled through no fault of the firer, the procedure outlined in *m* is followed.

p. More than five shots in rapid fire.—When a target has more than five hits in rapid fire, the target is not marked, except when all the hits have the same value, when the target is marked and the firer given that value for each shot fired by him.

q. Score cards and scoring.—(1) Entries on all score cards are made in ink or with indelible pencil. No alteration or correction is made on the card except by the organization commander, who initials each alteration or correction made.

(2) The scores at each firing point are kept by a noncommissioned officer of some organization other than that firing on the target to which he is assigned, except in case of a 1-company garrison when company officers exercise special care to insure correct scoring. As soon as a score is completed the score card is signed by the scorer, taken up and signed by the officer supervising the scoring, and turned over to the organization commander. Except when required for entering new scores on the range, score cards are retained in the personal possession of the organization commander and not allowed in the hands of an enlisted man from the beginning of record practice until the required reports of range practice have been rendered.

(3) In the pit the officer keeps the scores for the targets to which he is assigned. As soon as a score is completed he signs the score card. He turns these cards over to the organization commander at the end of the day's firing or at such times as requested.

(4) Upon completion of record firing and after the qualification order is issued the pit score cards of each man are attached to his official score card kept at the firing point. These cards are kept available for inspection among the company records for one year and then destroyed.

■ 60. COMPUTING SCORES.—The soldier's individual score is computed on a percentage basis. The soldier's percentage in firing each of the tables listed in paragraph 54 is calculated separately, then the sum of these percentages is divided by three to give the final average percentage.

■ 61. CLASSIFICATION.—*a.* The individual classification to be attained and the method of determining qualification are as prescribed in AR 775–10. Firers are classified as pistol experts, pistol sharpshooters, pistol marksmen, and unqualified.

AUTOMATIC PISTOL, CAL. .45, M1911 AND M1911A1   61–62

b. In case of a tie occurring at the dividing line between classifications, all who tie are given the higher classification and the percentage of the lower classification is reduced accordingly.

SECTION IV

KNOWN-DISTANCE TARGETS AND RANGES; RANGE PRECAUTIONS

■ 62. TARGETS.—a. *Target E.*—Target E is a drab silhouette about the height of a soldier in a kneeling position made of bookbinder's board or other similar material (fig. 14). Hits

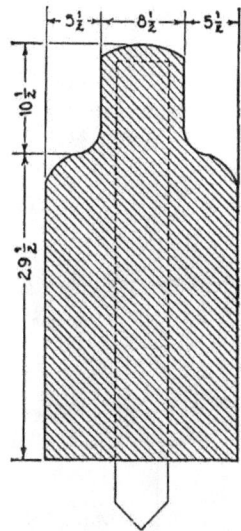

FIGURE 14.—Target E.

are valued at 1. Any shot cutting the edge of the target is a hit.

b. *Target E—bobbing.*—Target E—bobbing is so arranged as to be fully exposed to the firer for a limited time, edge of target toward firer when target is not exposed (fig. 15).

c. *Target L.*—Target L is a rectangle 6 feet high and 4 feet wide, with black circular bull's-eye 5 inches in diameter and seven outer rings (fig. 16). Value of hits in the bull's-eye, 10. The diameter of each ring and value of hits are as follows:

Diameter	Value of hit
8½ inches	9
12 inches	8
15½ inches	7
19 inches	6
22½ inches	5
26 inches	4
46 inches	3
Outer, remainder of target	2

d. Small-bore targets.—No specific targets are prescribed for small-bore practice with the pistol and any suitable targets may be used. As the targets specified in the table in section V below are not available, the following targets issued by the Ordnance Department are suitable and may be used: Rifle—SB–A–2, SB–A–3, and SB–B–5.

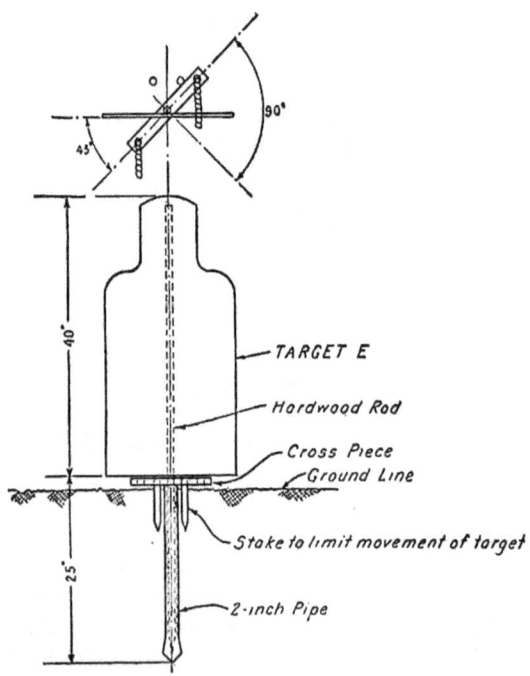

FIGURE 15.—Target E—bobbing.

AUTOMATIC PISTOL, CAL. .45, M1911 AND M1911A1   63

■ 63. RANGES.—*a. General.*—Class A target ranges for the rifle as described in FM 23–5 and 23–10 may be used for pistol practice, if available. The pistol target L may be placed on the sliding target carriage for both slow and rapid fire. Bobbing targets are not ordinarily placed in the pits of rifle ranges, but are set up nearby. If sufficient space is available, ground other than rifle ranges is used for pistol practice.

b. Rules for selection.—As the nature and extent of the ground available for pistol practice and also the general climatic conditions are often widely dissimilar for different military posts, it is impossible to prescribe any particular rules governing the selection of ranges, but only to express certain

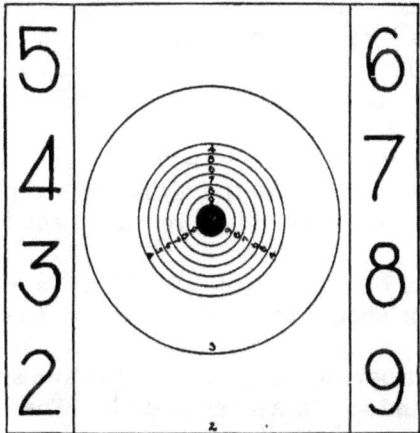

FIGURE 16.—Target L.

general conditions to which ranges should be made to conform.

c. Safety necessary.—For posts situated in thickly settled localities where the extent of military reservation is limited, the first condition to be fulfilled is that of safety for those living or working near, or passing by the range. This requirement can be secured by selecting ground where a natural butt is available or by making an artificial butt sufficiently extensive to stop wild shots. For complete safety there should be no road, building, or cultivated ground nearer than 300 yards to either flank of the range, nor than 1,600 yards to its rear.

d. Direction of the range.—If possible a range should be so located that the direction of firing is toward or slightly to the east of north. Such location gives a good light on the face of the targets during the greater part of the day. However, safety and suitable ground are more important than direction.

e. Best ground for range.—Smooth level ground or ground with only a very moderate slope is best adapted for a range. The target should be on the same level with the firer or only slightly above him. Firing down hill should be avoided.

f. Size of range.—The size of the range is determined by its plan and by the number of troops that will fire on it at a time.

■ 64. PRINCIPLES GOVERNING CONSTRUCTION.—*a. Intervals between targets.*—Intervals between targets are equal to the width of the targets themselves. When the necessity exists for as many targets as possible in a limited space this interval may be reduced. Bobbing targets should be placed a minimum of 5 yards apart.

b. Protection for markers.—When pits are not used, markers remain in rear of the firing line except during cessation of fire when their duties require them to move to the targets.

c. Artificial butts.—If an artificial butt is constructed as a bullet stop, it should be of earth not less than 30 feet high with a slope of not less than 45 degrees. It should extend about 5 yards beyond the outside targets and should be placed as close behind the targets as possible. The slope should be sodded.

d. Hills as butts.—A natural hill to form an effective butt should have a slope of not less than 45°. If originally more gradual it should be cut into steps, the face of each step having that slope.

e. Number of targets.—Each target should be designated by a number.

f. Measuring the range.—The range should be carefully measured and marked with a stake in front of each target at each firing point. The stakes should be about 12 inches above the ground and painted white. These stakes then designate the firing points for the different targets at the different distances. Particular care should be taken that each stake thus placed is at right angles to the face of its own target.

g. Danger signals.—One or more danger signals are placed near the range to warn passersby when firing is in progress. They should be placed on the roads or on the crest of the hill where they can be plainly seen by those passing.

Section V

SMALL-BORE PRACTICE

■ 65. General.—Where facilities and equipment permit, all soldiers who have satisfactorily passed the examination on preparatory exercises should be advanced to small-bore practice before taking up range practice. The actual firing and observed results stimulate endeavor. Rapid progress, particularly in teaching trigger squeeze, may be made. There is no recoil or loud report to induce nervcusness or flinching and the soldier soon learns that he can make good scores if he observes the proper methods and precautions in which he has been instructed. Small-bore practice is not only valuable to the beginner but it affords to the good shot a means of retaining his efficiency throughout the year. Continued practice is essential in maintaining pistol marksmanship.

■ 66. Object.—The object of small-bore practice is to provide a form of marksmanship training with the caliber .22 pistol and ammunition which represents the application of the principles taught in the preparatory exercises. Small-bore practice provides an excellent means of improving the shooting of organizations and sustaining interest in marksmanship throughout the year. Every effort should be made by all organizations to fire the small-bore course prior to the regular marksmanship season. The firing of this course enables the organization commander to visualize the state of training of his command and to concentrate his efforts on the training of those who are most deficient.

■ 67. Continuous Practice.—Small-bore practice should be carried on throughout the year subject to such limitations as may be imposed by available ammunition and range facilities. All persons who have never been properly instructed in shooting methods prescribed herein should be given a thorough course of preparatory instruction before being permitted to fire on the small-bore range. All small-bore practice is properly organized and supervised in accordance with the methods of instruction as prescribed in this manual.

■ 68. SMALL-BORE PRACTICE COURSE.—Prior to range firing the minimum number of scores shown in the following small-bore practice table should be fired by each individual required to fire the pistol, dismounted. The procedure of firing is similar to that in range firing. No reports of the results of small-bore practice are required but the firing record of individuals should be posted in order to stimulate interest and competition among the men of the organizations.

Small-bore practice table

Range	Slow-fire target		Rapid-fire target		Quick-fire target	
	Iron small-bore target or paper target X		Same as slow fire		Target E—bobbing	
	Time	Scores	Time	Scores	Time	Scores
5 yards	No time limit	2	15 seconds per score.	2		
10 yards	...do...	2	...do...	2		
15 yards					2 seconds per shot.	2

■ 69. ADDITIONAL PRACTICE.—In addition to the minimum number of scores prescribed in the table above, small-bore practice should be carried on throughout the year, the amount and details of the practice being left to the discretion of the organization commander. Varied targets such as tin cans, bottles, pendulums, and moving targets, stimulate interest. Matches between individuals and teams of the same or different units should be promoted.

CHAPTER 4

MARKSMANSHIP, KNOWN-DISTANCE TARGETS, MOUNTED

	Paragraphs
SECTION I. Preparatory training	70–72
II. Courses to be fired	73–74
III. Conduct of range practice	75–78
IV. Known-distance targets and ranges; range precautions	79–81
V. Small-bore practice	82–84

SECTION I

PREPARATORY TRAINING

■ 70. INSTRUCTION THROUGHOUT THE YEAR.—Mounted exercises with the pistol will be carried on during the entire year. By exercises is meant simulated firing at different objects at all gaits. In practicing pistol attacks in field exercises the troopers are required to observe the proper form in simulating firing.

■ 71. SIMULATED FIRING EXERCISES.—With the targets set up as required for actual firing the trooper simulates fire to the right front, right, right rear, left front, and left, bringing his horse to a halt upon the completion of each run. For firing mounted, it is of special importance that the pistol be pointed in prolongation of the forearm, and not to the right of the forearm, as is the common tendency. To enable men with average sized or smaller hands to avoid this tendency the pistol barrel is swung well to the left even though only the tip of the index finger will then reach the trigger. These exercises should be made as realistic as possible, as a perfunctory execution of them is a waste of time.

  a. The trooper is given practice in simulating fire to the right (left) at the WALK and the GALLOP. From RAISE PISTOL the trooper points his pistol by thrusting it toward the target by the shortest possible line, not by flourishing the weapon over the arc of a circle. This movement should be steady, easy, and smooth. In bringing the pistol downward he lowers the forearm in the direction of the target, keeping the elbow slightly bent, the forearm and pistol in the same straight line, and the wrist rigid. He then leans his body well toward the target, resting the bridle hand on the horse's crest, at the same time forcing the right elbow well to the left, thus getting the elbow in the same vertical plane as the wrist and shoulder.

FIGURE 17.—Beginning the thrust.

He then extends the forearm so as to thrust toward the target, keeping the elbow forced well to the left. There are thus four movements; leaning the body toward the target, lowering forearm and pistol, forcing elbow to the left, and a thrusting toward the target. Throughout the exercise he keeps his eyes fixed on the center of the target and concentrates his mind on making hits (figs. 17, 18, and 19).

AUTOMATIC PISTOL, CAL. .45, M1911 AND M1911A1   71

b. The trigger squeeze is started at the beginning of the thrust and is continued so that when the arm is fully extended or the thrust completed, the squeeze is completed and the piece is fired. Upon completion of the thrust the target should be seen along the top of the pistol. If the next target is on the same side as the one just fired on, the body remains leaning well toward the target and the elbow is bent ready

FIGURE 18.—Thrust continued.

for the next thrust. During the processes of firing, the bridle hand remains on the horse's crest.

c. In firing to the front (fig. 20), the trooper pushes his horse forward into the extended gallop and with reins gathered up short, leans well forward and extends his pistol arm as far as possible to the front. To accomplish this, he should half stand in the stirrups. When the target is very low or

69

close, the trooper extends his arm well to the front, being careful to have the muzzle of the pistol barrel well in front of the horse's face. He selects his target, fixes his eyes upon it with concentrated effort, catches sight of the target over his pistol barrel, and squeezes his trigger with steadiness and precision. He remains in this attitude simulating the firing of successive shots until he has passed the targets.

d. In firing to the rear (fig. 21), the trooper applies the principles of arm extension, seeing the target over the pistol and squeezing the trigger as described in the preceding exer-

Figure 19.—Completed thrust.

cises, but instead of half rising in the stirrups and leaning forward, he sits well down in the saddle and rotates his body to the right and rear, being careful not to disturb his bridle hand nor to allow his lower legs to fly to the front. One of the principal causes for missing the target is failure of the firer to keep his eyes on the target until *after the shot is fired.* There is a tendency for the firer to become concerned with the next target with the result that he takes his eyes off the target he is engaging an instant before the shot is fired. The firer should prepare to engage the next target only *after* the shot is fired at the target being engaged.

■ 72. TRAINING HORSES.—Horses should be accustomed to the sight of barriers and targets and to the sound of firing. Targets placed in the corral and targets and barriers placed near the road to the drill ground where the horses pass

FIGURE 20.—Firing to the front.

them frequently soon familiarize them with the odd shapes and colors. A horse that shows fear of a target should not be forced up to it roughly but should be ridden close to the targets gradually and patiently. A practicable way of inuring horses to the sound of firing is to locate them on a picket line or otherwise in close proximity to the firing points

FIGURE 21.—Firing to the rear.

FIGURE 22.—Firing to the left.

during rifle and dismounted pistol practice. The training of horses to accustom them to the noise of firing should not be hurried. Calmness usually develops gradually. Horses should always be exercised before firing in order to calm them. A group of horses stand firing more quietly than an individual horse. The most nervous horses should have quiet horses on either side during training. As they grow steadier, firing by individuals armed with pistol and blank ammunition should be done at drill in line formations and at increased gaits.

Section II

COURSES TO BE FIRED

■ 73. INSTRUCTION PRACTICE.—The following table indicates the minimum amount of firing on the range during instruction practice:

Table I

	Runs twice around the course	Targets	Range	Gait	Time limit
Firing to the right front, right, right rear, left front, and left.	Minimum of 2 runs of 14 shots each.	4 M 3 E	5 to 7 yards	Gallop	52 seconds for each run, twice around the course.

■ 74. RECORD PRACTICE.—The following table indicates the firing to be done during record practice:

Table II

	Runs twice around the course	Targets	Range	Gait	Time limit
Firing to the right front, right, right rear, left front, and left.	2 runs of 14 shots	4 M 3 E	5 to 7 yards	Gallop	52 seconds for each run, twice around the course.

Section III

CONDUCT OF RANGE PRACTICE

■ 75. RANGE FIRING.—*a. Marking, scoring, etc.*—The targets are designated Nos. 1, 2, etc., as shown in figure 23. The number of men detailed as markers is left to the discretion of the officer conducting the firing. The markers take position approximately opposite their targets and on the designated line. When a complete score has been fired, the markers run to their targets, examine them, face the scorer, and call in numerical order the hits or misses, as for example, "No. 1, a hit," "No. 2, a miss," "No. 3, 2 hits," etc. After calling the hits or misses, the markers cover any shot holes with pasters and run back to their positions. One noncommissioned officer detailed as scorer is posted behind the markers at a convenient place to hear the calls. The regulations for scoring and recording scores outlined in section III, chapter 3, govern in mounted pistol fire as far as applicable.

b. Value of hits.—Each hit on a target counts one point.

c. Defective cartridges and malfunctions.—(1) If a defective cartridge or a malfunction of the pistol causes failure to fire at any one of the targets from numbers 1 to 7, inclusive, during the first time around the course, the firer returns to the starting point at RAISED PISTOL where he corrects the defect. The pistol is then loaded so as to provide one cartridge for each target numbered 1 to 7 at which he failed to fire. The firer then starts his run anew, firing on those targets only which he has previously failed to fire upon.

(2) If a malfunction occurs between targets numbers 7 and 1, the firer returns to the starting point at RAISED PISTOL, where he corrects the defect, places an empty magazine in the pistol and draws the slide back. The firer then starts his run anew. After passing target No. 7, he withdraws the empty magazine, inserts a loaded magazine, and commences firing at target No. 1.

(3) If a defective cartridge or malfunction of the pistol causes a failure to fire at any of the targets from numbers 1 to 7, inclusive, during the second time around the course, the firer continues around the track to the starting point at RAISED PISTOL, where he corrects the defect. The pistol is then loaded so as to provide one cartridge for each target

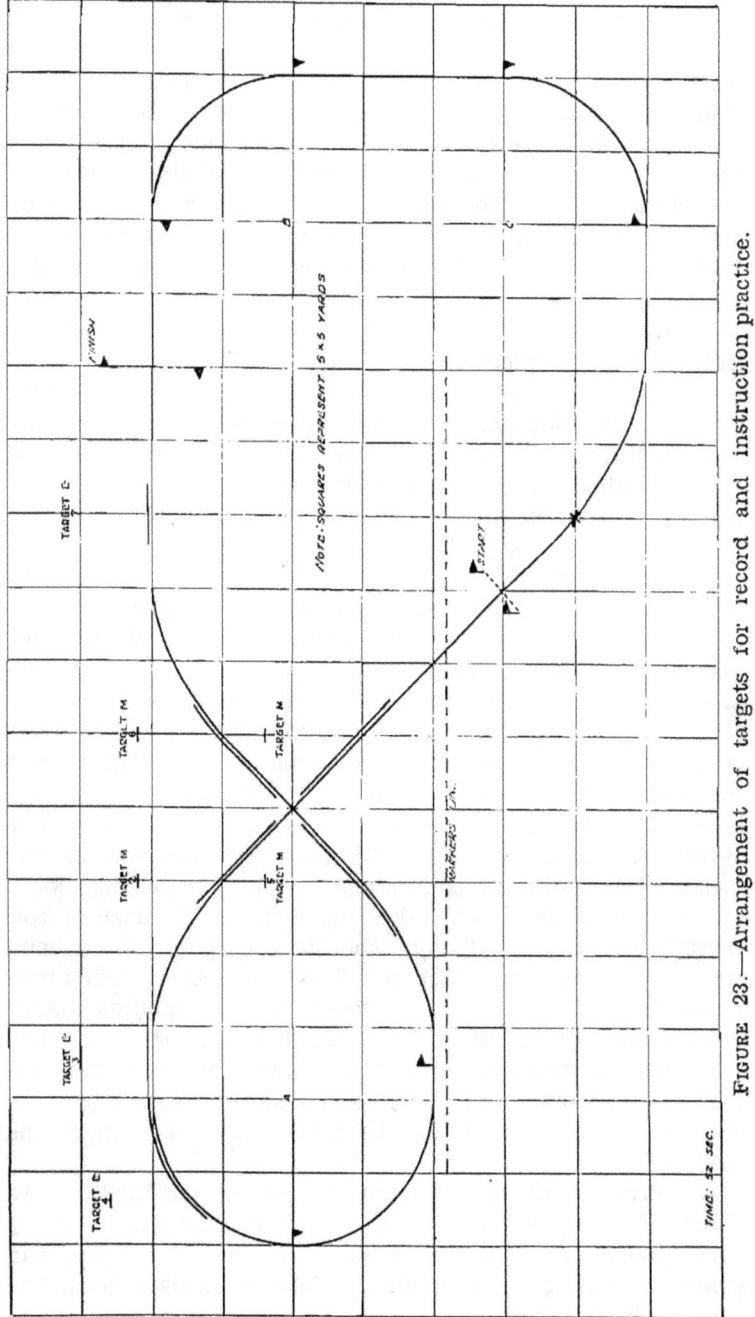

FIGURE 23.—Arrangement of targets for record and instruction practice.

numbers 1 to 7 at which he failed to fire the second time. The firer then starts his run anew, firing at those targets only from numbers 1 to 7, inclusive, at which he previously failed to fire during the second time around the course.

d. Time limit.—The time limit for each run is 52 seconds. The time is measured from the instant the firer passes the flag at the starting point to the instant he passes the flag at the finish point. In case, due to a defective cartridge or malfunction, the firer is permitted to renew his run, the time starts as prescribed for a regular run.

e. Horses.—So far as practicable, individual firers are required to ride the horses regularly assigned to them. In no case in the Regular Army is any horse used by more than two individuals, nor are more than 25 percent of the individuals of a troop permitted to fire from horses other than those regularly assigned to them.

f. Instruction practice.—(1) The firer is equipped with pistol, lanyard, field belt, and two loaded magazines of seven rounds each. Just prior to starting a run he inserts a loaded magazine, loads and locks the pistol, and takes the position of RAISED PISTOL. The other magazine remains in the magazine pocket. Starting with the gallop at X, the firer rides past the starting point, unlocks his pistol, and rides around the course as indicated in figure 23, firing one shot at each one of the targets numbered 1 to 7 as he encounters them. When he has fired his seventh shot the firer, continuing around the course at the gallop, changes magazines, loads and locks the pistol, and takes the position of RAISED PISTOL. Continuing past the flag at the starting point the second time he unlocks his pistol and fires again at targets numbered 1 to 7 as each is encountered. Upon changing magazines after passing target No. 7 the first time, the firer may place the empty magazine in any secure place most convenient to him such as a shirt pocket, inside the shirt, or if he so desires, he may return the magazine to the magazine pocket.

(2) Prior to making a complete run as outlined in (1) above the troopers should be afforded practice in going through the course and simulating fire on all targets, and going through the course firing blank cartridges from the caliber .45 pistol at six targets selected at random.

(3) To accustom horses and riders to the type of firing and to minimize the chances of accidents, the first run of instruction practice may be divided into two phases as follows:

(a) In this first phase the soldier fires ball cartridges at targets numbers 1 to 7, removes the empty magazine, inserts another empty magazine while en route to target No. 1, and simulates fire on targets numbered 1 to 7 during the second time around the course.

(b) In the second phase the soldier, having an empty magazine in his pistol and the slide held back by the slide stop, simulates fire on targets numbered 1 to 7, removes the empty magazine, inserts a loaded magazine while en route to target No. 1, and fires ball cartridges on targets numbered 1 to 7 the second time around the course. The slide is held back at the beginning of the run in order that as the firer passes target No. 7 the slide will be in the position it normally would be in if actual firing had taken place during the first time around the course.

(c) Other variations, such as loading less than seven rounds in a clip and firing at certain designated targets, may be prescribed by the organization commander.

(4) Penalties are exacted as follows:

(a) For each 5 seconds or fraction thereof in excess of the prescribed time limit of 52 seconds, one point is deducted from the final score.

(b) For each instance during a run that the horse assumes a gait other than the gallop, one point is deducted from the firer's score. This penalty is not to be construed as applicable to a case where the horse in changing leads hits two or three beats of the trot.

g. Record practice.—(1) The individual record practice is fired in the same manner and with the same penalties as prescribed in *f* (1) and (4) above.

(2) A second continuous run completes the record practice.

■ 76. COMPUTING SCORES.—The soldier's individual score is based upon the total number of hits.

■ 77. CLASSIFICATION.—The individual classification to be attained and the minimum aggregate scores required for qualifications are as prescribed in AR 775-10. Firers are classified as mounted pistol experts, pistol sharpshooters, pistol marksmen, and unqualified.

■ 78. SAFETY PRECAUTIONS ON THE RANGE.—*a. Safety precautions.*—In order to teach the trooper to handle his pistol with safety he is frequently required to go through the motions of loading and unloading, locking and unlocking the pistol while mounted, both at the halt and at all gaits. The troopers should also be practiced in withdrawing and inserting magazines at all gaits. During this preparatory instruction the trooper must be taught by exercises in simulated fire to observe the following rules: (This will be done daily until the proper movements in their proper sequence become matters of habit and are instinctively performed when they become necessary.)

(1) Always lock the pistol after loading and keep it locked until about to fire.

(2) When a cocked pistol is held in the hand it must always be held at RAISE PISTOL until it is locked, or until it is necessary to fire, load, or unload. If it becomes necessary to lower the pistol for any other purpose, the pistol must first be locked. Mounted men should never under any circumstances use both hands on the reins when the pistol is drawn.

(3) At the slightest misbehavior of the horse, the pistol must be locked. The trooper then makes a fresh start to complete his firing. If the horse rears, plunges, bolts, stumbles, or leaves the track, the trooper should instinctively lock the pistol.

b. Safety exercises.—In order to habituate the troopers to obey these rules the following exercises must be practiced frequently:

(1) The trooper, at RAISE PISTOL and with pistol cocked and locked, takes the track in front of the targets and unlocks the pistol. He then rides alternately off the track and onto it again, each time locking pistol as he leaves and unlocking pistol as he returns. This exercise is executed at all gaits. It simulates a misbehaving horse and teaches the trooper to lock pistol at once at the beginning of such misbehavior.

(2) The trooper, with pistol cocked and locked, takes the track at RAISE PISTOL. He now unlocks the pistol as if about

to fire, and allows his bridle hand to slide along the reins until the latter are too long; whereupon he locks and returns pistol, and uses his pistol hand to assist in readjusting the reins. He next draws and raises pistol, continuing the exercise until he has passed the end of the track. This exercise is practiced at all gaits.

(3) Exercises for loading, unloading, withdrawing, and inserting magazine must be improvised and practiced at all gaits. Practice in the proper method of returning and withdrawing pistol should be emphasized. In each of these operations the flap of the holster must be held back by the heel of the hand while the pistol barrel is pointing downward and slightly away from the horse. Under no circumstances should the flap be raised by the muzzle of the pistol or the pistol barrel pointed in toward the leg or the horse. The finger should be kept outside the trigger guard and care must be exercised to insure that the piece does not become unlocked during these operations.

SECTION IV

KNOWN-DISTANCE TARGETS AND RANGES; RANGE PRECAUTIONS

■ 79. TARGET E.—Target E is a drab silhouette, representing a kneeling figure, made of book binders board or similar material. (See fig. 14.)

■ 80. TARGET M.—Target M consists of two parts: The upper is Target E and the lower is a trapezoidal piece whose upper edge is placed closely against the lower edge of Target E. It is made of material similar to Target E. (See fig. 24.)

■ 81. CONSTRUCTION OF THE COURSE.—The course is laid out as shown in figure 23. The length of the course measured from start to finish is 356 yards. The area should be level or with but a gentle slope. Points on the course are marked with flags or posts as indicated on the diagram. The curves at A, B, and C have a 10-yard radius. The other curves have a radius of about 20 yards. Targets are of the E and the M type and are located as shown by the numbers 1 to 7, figure 23. Barriers at least 18 inches in height are placed as indicated by the heavy lines in figure 23. A flag is placed at the starting point and at the finish line as indicated in figure 23.

SECTION V

SMALL-BORE PRACTICE

■ 82. GENERAL.—The general matter contained in section V, chapter 3, applies equally to mounted pistol training. Mounted pistol practice with the caliber .22 pistol is of great value in accustoming the horses to firing from the saddle and to running the mounted course, as the lesser noise and

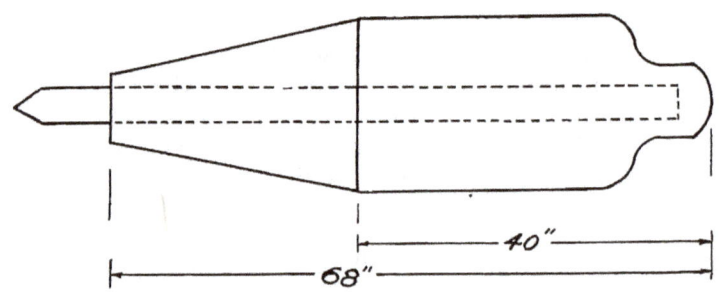

FIGURE 24.—Target M.

concussion does not distract them from their work as does the caliber .45 pistol. The lesser cost of ammunition permits more instruction practice than with caliber .45. Organizations have been known to qualify 100 percent of their men at mounted pistol practice by conducting all of their instruction practice with the caliber .22 and using the caliber .45 for the first time each season in record practice.

■ 83. SMALL-BORE RANGES.—*a.* The range used for small-bore mounted pistol practice is the same as described for the qualification course. (See par. 81.)

b. The range described in paragraph 91 may be used for mounted firing at field targets with small-bore pistols at the discretion of the organization commander.

■ 84. SMALL-BORE COURSES.—*a.* The course fired is the same as that described for instruction and record practice, the qualification course (pars. 73 and 74).

b. In case small-bore pistols are used for mounted firing at field targets, the course is as prescribed in paragraph 92.

CHAPTER 5

FIRING AT FIELD TARGETS

	Paragraphs
SECTION I. Dismounted	85–88
II. Mounted	89–93

SECTION I

DISMOUNTED

■ 85. GENERAL.—*a.* After individual practice is completed, firing at field targets may be commenced by all organizations authorized to fire this practice.

b. This firing is practiced by squads.

■ 86. FIELD FIRING RANGE.—Eight targets E—bobbing are placed approximately in line with 5-yard intervals and with ropes sufficiently long to reach behind the starting line. The starting line is 50 yards from the targets and is marked by a flag at each side of the range. The area should be clear of trees and high brush, level or sloping gently upward toward the targets.

■ 87. FIELD FIRING COURSE.—The following table outlines the course to be fired. The course may be repeated at will as often as is consistent with ammunition allowances.

Formation	Scores	Targets	Range	Pace
Squad of 8 men deployed at 5-yard intervals.	2 scores, each man firing 7 shots in each score.	8 targets E—bobbing.	Between 50 and 15 yards.	Walk.

NOTE.—Hits are valued at 1. Any shot cutting the edge of the target is a hit.

■ 88. METHOD OF FIRING THE COURSE.—*a.* The squad is formed at the starting line facing the targets with 5-yard intervals. Pistols, unloaded but containing a full magazine, are in holsters.

b. The officer in charge of firing commands: 1. FORWARD, 2. MARCH. All men move toward the targets at the walk. By signal the officer in charge of firing causes the targets to be exposed for 6 seconds when the line has moved forward about 5 yards.

c. When the targets first appear the men individually draw and raise pistol and load. They fire as many shots as they desire at each exposure of the target. During the periods when the targets are exposed all men must halt. During periods when the targets are not exposed all men move toward the targets at the walk, with pistols at RAISE PISTOL.

d. The officer in charge of firing causes the targets to be exposed 3 seconds after the line of men has advanced each 5 yards, until men reach a line 15 yards from the targets where the targets are exposed for the last time. Thus targets are exposed seven times, the first time for 6 seconds and all other times for 3 seconds.

e. Each man must fire only on the target directly in front of him.

f. Care must be taken that the line be kept approximately dressed. If a man gets 3 yards in front of, or behind the line, he must be cautioned of his error and must correct it immediately.

g. For purposes of safety a noncommissioned officer is stationed in rear of the squad and follows it forward during each run to enforce the provisions of *f* above.

h. Firing ceases on command of the officer in charge of firing. Men stand fast, pistols are unloaded, and the officer in charge of firing commands: 1. INSPECTION, 2. PISTOL, 3. RETURN, 4. PISTOL.

Section II

MOUNTED

■ 89. GENERAL.—*a.* After dismounted field firing is completed, mounted field firing is commenced by all cavalry rifle troops.

b. The restrictions regarding the use of horses contained in paragraph 75*e* do not apply to mounted field firing. Prior to the commencement of this practice, the commanding officer of each rifle troop selects from among the horses permanently assigned his troop those best trained for pistol firing. The horses so selected are used in field firing practice.

c. This firing is practiced by squads. Men not belonging to permanent squads are either formed into squads or are used to replace absentees in permanent squads.

d. A suitable marker is placed on each side of the course 35 yards from the line of targets to indicate the line at which firing may commence. This line is known as the firing line.

e. After crossing the firing line, any trooper who is passed by the trooper on his right or left ceases firing, locks his pistol, and takes the position of RAISE PISTOL.

f. A suitable marker is placed on each side of the course 5 yards in front of the line of targets to indicate a line beyond which there will be no firing. This line is known as the safety line.

g. When the squad reaches the safety line, or at any time in case of emergency, the whistle signal "Suspend firing" will be sounded. At this signal, all firing ceases, pistols are locked, and the position of RAISE PISTOL is taken.

h. In these exercises the squad leader is the guide upon whom the troopers maintain their alinement and preserve their intervals.

i. After passing beyond the targets, the squad leader halts the squad and gives the necessary commands for unloading and returning pistols. He then marches the squad off at a slow gait.

■ 90. PRELIMINARY EXERCISES.—As preliminary training for mounted field firing the following exercises are of value and should be repeated until horses are going smoothly and the instructor is assured that all members of the squad are thoroughly acquainted with the procedure.

a. Using the range lay-out described in paragraph 92a:

(1) The individual troopers make two or more runs with empty pistols, simulating fire. Runs are then made with the trooper firing ball ammunition, first with only one round and later with three rounds.

(2) The squad makes two or more runs with empty pistols, simulating fire. Runs are then made with the trooper on the right firing three rounds of ball ammunition. The squad

is formed in close order so that the firer is opposite the target on the right. All troopers except the firer simulate fire. Otherwise the run is conducted as prescribed in paragraph 92. By changing positions in the squad each trooper is permitted to fire.

b. Using the range lay-out described in paragraph 93*a* the exercises outlined in *a* above are practiced.

c. The amount of firing during preliminary exercises and field firing depends to a great extent upon ammunition allowances.

■ 91. FIELD FIRING COURSE (MOUNTED).—The following table outlines the course to be fired:

Formation	Scores	Targets	Range	Gait
Exercise No. 1: Squad of 8 troopers in close-order line.	3 runs, each trooper firing 3 shots in each run.	8 targets E, suspended.	Between 35 and 5 yards.	Extended gallop.
Exercise No. 2: Squad of 8 troopers deployed as foragers at a minimum of 5-yard intervals.	....do..........	9 targets M..	....do......	do.

NOTE.—The object of exercise No. 1 is to teach the men to fire to the front over their horses' heads while the squad is in line in close order. The object of exercise No. 2 is to teach the men to fire to the front at low targets while the squad is in line in extended order. Hits are valued at 1. Any shot cutting the edge of the target is a hit.

■ 92. EXERCISE NO. 1.—*a.* Eight silhouette targets E with 1½ yards between centers are suspended from a rope (or wire cable). The rope is stretched between poles erected 30 yards apart. The targets are suspended from the rope by any suitable means in such a manner that the lower edges of the targets when suspended will be about 10 feet from the ground.

b. The squad is formed in line in close order facing the targets and at not less than 75 yards distance from them. The squad leader, posted as No. 1 of the left four, requires each trooper to load his magazine with three cartridges. He causes the squad to LOAD and RAISE PISTOL, and then gives the necessary commands for putting the squad in motion and for increasing the gait to a GALLOP. The rate of the gallop is gradually increased until the squad is moving at approximately 16 miles per hour when it reaches the firing line. When the squad reaches this line the leader commences firing; this is the signal for the other troopers to commence firing. Each trooper leans well forward and fires carefully and deliberately at the target directly in front of him. Firing ceases at the whistle signal, "Suspend firing".

■ 93. EXERCISE No. 2.—*a.* Nine silhouette targets M with a minimum interval of 5 yards between centers are erected in line.

b. The squad is formed in line in close order facing the targets at not less than 100 yards distance from them. The squad leader, posted as No. 1 of the left four, requires each trooper to load his magazine with three cartridges. He causes the squad to execute LOAD and RAISE PISTOL, and then gives the necessary commands for putting the squad in motion, for deploying it, and for increasing the gait to a GALLOP. The rate of the gallop is gradually increased until the squad is moving at approximately 16 miles per hour when it reaches the firing line. When the squad reaches this line, the leader commences firing; this is the signal for the other troopers to commence firing. Each trooper leans well forward and fires carefully and deliberately at the target to the right of the interval through which he is to pass. Firing ceases at the whistle signal, "Suspend firing."

CHAPTER 6

ADVICE TO INSTRUCTORS

	Paragraphs
SECTION I. General	94– 95
II. Mechanical training	96–101
III. Manual of the pistol	102
IV. Marksmanship	103–106

SECTION I

GENERAL

■ 94. PROVISIONS NOT MANDATORY.—The information and suggestions contained in this chapter are not mandatory unless so specifically stated. They are furnished as a guide for the personnel responsible for the instruction of troops in the subjects contained herein.

■ 95. METHOD OF INSTRUCTION.—The applicatory system of instruction is used for instruction in subjects of the nature found in this manual. This system consists of explanation, demonstration, application (practical work), and examination.

 a. Explanation.—The initial explanation and demonstration of any particular phase of the instruction is presented to the assembled unit by the instructor assisted by essential demonstration personnel. The general purpose of the entire course or period of instruction should be explained first. The various phases or steps of the course should then be presented in a series of explanations and demonstrations.

 b. Demonstration.—(1) Demonstrations which are skillfully conceived and executed expedite and simplify instruction as well as stimulate interest. Successful demonstrations are usually short and concise. They leave the student with an exact impression stripped of superfluous details. The demonstrations incident to all subjects should be arranged in progressive sequence, and where practicable should alternate with practical work to permit the student to fix these successive phases of instruction in his mind.

 (2) The men who constitute the demonstration unit should be carefully selected for their intelligence, ability, and appear-

ance. They should be thoroughly trained and rehearsed in the duties they are to perform so that the demonstration will proceed smoothly and illustrate clearly and simply the phase of instruction being presented.

(3) The equipment used for demonstrations should be the best available. A demonstration platform or an area in which the students can be assembled quickly at a position from which they can see and hear every part of the demonstration is essential.

(4) Interest is added and valuable instruction given by repeating demonstrations, including common errors, and requiring the students to detect these errors.

  c. Application (practical work).—(1) This third step of instruction is of major importance since it gives the student an opportunity to actually accomplish that which has been previously explained and demonstrated.

(2) During the practical work phase of instruction best results are obtained if the unit is divided into groups. Groups should consist of from four to eight men depending upon the number of men undergoing instruction and the number of assistant instructors available. Each group is provided with a set of equipment and placed under the direct supervision of a trained assistant instructor. The group then executes the previously demonstrated phase of instruction, individuals rotating within the groups, until all men have mastered the instruction.

(3) The initial allotment of time and equipment should be made carefully. However, the instructor should not hesitate to alter this allotment if the majority of the men fail to master the instruction within the allotted time or are kept at one exercise to the point of boredom. The frequent rotation of duties within each group is preferable to keeping each man in one position for a long time.

  d. Examination.—An informal oral or practical examination should be conducted upon completion of each phase of instruction. In addition to the required examination before starting range practice, the organization commander should conduct such additional examinations as are necessary to insure that all men have completed the training.

Section II

MECHANICAL TRAINING

■ 96. GENERAL.—The entire unit to be instructed is assembled in a suitable area and divided into conveniently sized groups, each under the supervision of an assistant instructor. The instruction is centralized under the supervision of the unit instructor. Explanation and demonstration are concurrent, each assistant instructor demonstrating the elements of the particular phase of instruction as the instructor explains it from the platform. For short periods of practical work the instruction is decentralized under the assistant instructors.

■ 97. DISASSEMBLY AND ASSEMBLY OF THE PISTOL.—*a. Equipment required.*—One pistol with magazine per man, one pistol cleaning kit per group.

b. Procedure.—(1) Have assistant instructor disassemble and assemble the pistol while the instructor is explaining the procedure.

(2) *Practical work.*—Assistant instructors explain and demonstrate the procedure, and each student performs each operation in unison with the assistant instructor. When acquainted with the procedure each student disassembles and assembles the pistol without assistance.

(3) Ask questions.

■ 98. CARE AND CLEANING.—*a. Equipment.*—Same as paragraph 97 plus additional for special demonstrations.

b. Procedure.—(1) Explain the need of keeping the pistol clean, lubricated, and in proper condition, comparing it with any other piece of machinery.

(2) Explain the proper method of cleaning, at the same time demonstrating the process.

(3) Instruction in cleaning is carried out throughout the year under supervision of squad and platoon commanders.

(4) Ask questions. Inspect pistols frequently.

■ 99. FUNCTIONING.—*a. Equipment.*—Same as paragraph 97.

b. Procedure.—(1) Explain the various steps which take place in loading, unloading, and firing the pistol.

(2) Students practice loading a magazine, inserting the magazine in the pistol, and loading and unloading the pistol.

(3) Ask questions.

■ 100. ACCESSORIES—*a. Equipment.*—One each of the accessories listed in paragraph 16.

b. Procedure.—(1) Explain and demonstrate the use of each accessory.

(2) Students examine accessories.

(3) Ask questions.

■ 101. INDIVIDUAL SAFETY PRECAUTIONS.—*a. Equipment.*—Same as paragraph 97.

b. Procedure.—(1) Explain and demonstrate the various safety rules listed in paragraph 25. Explain and demonstrate tests for safety devices.

(2) Students practice the application of the various safety rules, and make the tests of safety devices. Observation of the rules for safety becomes a habit only after constant practice over a long period. Squad, platoon, and other commanders should be constantly alert to enforce the safety rules at every opportunity.

(3) Ask questions.

SECTION III

MANUAL OF THE PISTOL

■ 102. GENERAL.—*a.* Instruction in the manual of the pistol is carried out concurrently with dismounted and mounted drill and with previous instruction in this chapter.

b. Manual of the pistol lends itself well to the applicatory system of instruction.

c. Equipment.—Each man is equipped with a pistol with magazine, holster, belt, magazine pocket, two extra magazines, and pistol lanyard (for mounted instruction).

d. Procedure.—(1) The instructor explains and demonstrates each movement in the manual of the pistol, employing a trained demonstration unit.

(2) Groups are separated under assistant instructors and are drilled in the various movements.

(3) Each group is tested by the instructor at the end of each period and a critique is conducted.

Section IV

MARKSMANSHIP

■ 103. GENERAL.—*a.* Marksmanship is the basic step in training the soldier to employ successfully the pistol in combat. A soldier will subconsciously employ in combat the principles he has been taught in marksmanship, hence these principles must be sound.

b. The procedure used in conducting marksmanship instruction is similar to that used in the preceding sections of this chapter except that it is more decentralized. During instruction in preparatory exercises the entire unit is assembled initially under the unit instructor assisted by a trained demonstration unit. Following the initial explanation and demonstration the groups move to their individual sets of equipment and start practical work under the assistant instructors.

c. Firing exercises should be conducted under centralized control.

■ 104. PREPARATORY RANGE TRAINING.—*a. General.*—(1) A thorough course in preparatory range training is essential. During this period the soldier learns all the mechanics of target practice except actual firing. Preparatory training may be done in barracks or other nonfiring areas.

(2) Adequate time should be allowed and thorough supervision provided to insure that each man has thoroughly mastered the instruction before he is permitted to fire.

(3) Each step is taken up in proper order and training in that step completed by each man before the next step is begun. If men fail to progress uniformly, groups should be rearranged so that instruction will not be held up by men who are slow to learn.

(4) A careful record of the progress of each man and each group should be kept in order that the instructor will know the progress of instruction and when the men are ready for range practice.

b. Equipment per group.—(1) One sighting bar.
(2) One pistol rest.
(3) Two small aiming disks.
(4) One 5-inch aiming disk.

(5) Two small boxes with paper tacked on one side.

(6) One target frame on which is placed a blank sheet of paper at least 2 feet square.

(7) One target L.

(8) One target E, bobbing.

(9) One pistol with magazine, holster, belt, magazine pocket, and two extra magazines per man.

(10) Material for blackening sights.

(11) Tissue paper for copying shot groups.

(12) Pencils.

(13) Additional equipment such as blackboard, charts, and drawings as decided by the instructor.

b. Procedure.—(1) Each phase of preliminary training is explained and demonstrated when instruction in that phase begins by the unit instructor, employing a trained demonstration group.

(2) Groups are separated and practical work is conducted under the supervision of assistant instructors.

(3) Examination is conducted by asking questions and by observing the results obtained by each man during practical work.

■ 105. FIRING OF COURSE.—*a. General.*—Details of administration and supply are determined to a large extent by the number of men undergoing instruction and the range facilities available. These matters should be anticipated in order that men who are firing will not be distracted. Men who are waiting to fire may be perfecting preliminary instruction or they may derive valuable instruction by watching others fire and by listening to critiques.

b. Equipment.—(1) *Pistols.*—Instruction in marksmanship is facilitated by having all pistols in perfect mechanical condition.

(2) *Magazines.*—Damaged or inoperative magazines are the greatest single cause for malfunctioning of the pistol. Principal defects in magazines are dents, spread lips, and the presence of sand or dirt below the follower.

■ 106. CONSTRUCTION OF TARGETS AND RANGES.—*a. General.*—For detailed information relative to targets and target accessories see Table of Allowances, Targets, and Target Equipment.

106 AUTOMATIC PISTOL, CAL. .45, M1911 AND M1911A1

b. Targets.—When regular printed targets are not available, suitable substitutes can be made on sheets of wrapping paper. Dimensions should be accurate. Improvised targets can be made in large numbers by improvising a stencil using heavy linoleum.

c. Ranges.—(1) *General.*—The range should be level and open and if practicable, so located that fire can be delivered against a steep hill or bank in rear of the targets. Semipermanent bases should be constructed to facilitate placing targets in position and changing targets with the minimum delay or confusion.

(2) *Dismounted course.*—Targets should be spaced from 3 to 5 yards apart. The depth of the range should be not less than 40 yards.

(3) *Mounted course.*—Dimensions outlined in paragraph 75 should be accurately observed.

(4) *Field firing.*—Ground rising gently toward the targets with a hill in rear of the targets should be selected if available. Avoid shooting down hill. The range should be at least 50 yards wide by 100 yards long.

INDEX

	Paragraph	Page
Accessories	16, 100	15, 89
Additional practice	69	66
Aiming (first step)	46	29
Ammunition:		
Ballistic data	24	18
Care, handling, and preservation	22	16
Classification	18	15
Grade	20	16
Identification	21	16
Lot number	19	15
Storage	23	18
Assembling	4	7
Assembly of pistol	97	88
Ballistic data on ammunition	24	18
Care of ammunition	22	16
Care of pistol	6, 98	8, 88
After firing	7	9
During cold weather	9	10
During gas attacks	10	10
On the range	8	10
Points to be observed	11	10
Chamber:		
To close	31	22
To open	30, 40	22, 24
Classification, individual	61, 77	60, 77
Classification of ammunition	18	15
Cleaning of pistol	6, 7, 11, 98	8, 9, 10, 88
Close chamber	31	22
Coaching methods	55	52
Computing scores	60, 76	60, 77
Construction of—		
Course	81	79
Targets and ranges	106	91
Principles governing	64	64
Continuous small-bore practice	67	65
Course:		
Construction	81	79
Field firing	87	81
Mounted	91	84
Firing of	105	91
Method	88	81
Courses, small-bore practice	68, 84	66, 80
Data on pistols	2	4
Description of pistols	1	1
Detailed functioning	14	12
Details, target	58	57
Disassembling	3	5
Disassembling of pistol	97	88
Examination (sixth step)	51	46
Exercise—		
No. 1	92	84
No. 2	93	85

INDEX

Exercises:	Paragraph	Page
Preliminary	90	83
Simulated firing	71	67
Field firing—		
Course	87	81
Mounted	91	84
Range	86	81
Fifth step (quick fire)	50	45
Fire pistol	37	24
Firing, range	75	74
Firing exercises, simulated	71	67
Firing of course	105	91
First step (aiming)	46	29
Fourth step (rapid fire)	49	42
Functioning of pistol	14, 99	12, 88
General data	2	4
Grade of ammunition	20	16
Handling of ammunition	22	16
Horses, training	72	71
Identification of ammunition	21	16
Individual safety precautions	101	89
Insert magazine	32, 41	22, 24
Inspect pistol	35, 44	23, 25
Instruction:		
Method	95	86
Practice	53, 73	50, 73
Preparatory training	45, 70	26, 67
Load pistol	33, 42	22, 25
Lot number of ammunition	19	15
Magazine:		
To insert	32, 41	22, 24
To withdraw	29, 39	22, 24
Manual of the pistol, instruction in	102	89
Method of—		
Instruction	95	86
Operation	12	11
Methods, coaching	55	52
Object of small-bore practice	66	65
Open chamber	30, 40	22, 24
Operation, method	12	11
Organization, range	57	56
Points to be observed in care of pistol	11	10
Pistol:		
Care and cleaning	6–11, 98	8, 10, 88
Description	1	1
Disassembly and assembly	3, 97	5, 88
Functioning of	14, 99	12, 88
General data on	2	4
To fire	37	24
To inspect	35, 44	23, 25
To load	33, 42	22, 25
To raise	28	22
To return	36	24
To unload	34, 43	23, 25

INDEX

	Paragraph	Page
Position (second step)	47	35
Practice:		
Instruction	53, 73	50, 73
Preparatory training	45	26
Record	54, 74	51, 73
Preliminary exercises	90	83
Preparatory range training	104	90
Preservation of ammunition	22	16
Principles governing construction of targets and ranges	64	64
Provisions not mandatory	94	86
Quick fire (fifth step)	50	45
Raise pistol	28	22
Range:		
Firing	75	74
Field	86	81
Safety precautions on	56, 78	56, 78
Training, preparatory	104	90
Organization	57	56
Ranges	63	63
Construction	106	91
Principles governing	64	64
Small-bore	83	80
Rapid fire (fourth step)	49	42
Record practice	54, 74	51, 73
Regulations governing	59	58
Return pistol	36	24
Rules for safety	25	18
Safety devices	13	12
Safety precautions:		
Individual	101	89
On the range	56, 78	56, 78
Safety rules	25	18
Safety tests	26	20
Scores, computing	60, 76	60, 77
Second step (position)	47	35
Simulated firing exercises	71	67
Sixth step (examination)	51	46
Small-bore practice:		
Additional	69	66
Continuous	67	65
Courses	68, 84	66, 80
Object	66	65
Ranges	83	80
Spare parts	15	14
Storage of ammunition	23	18
Target details	58	57
Target E	79	79
Target M	80	79
Targets	62	61
Construction	106	91
Principles governing	64	64

INDEX

	Paragraph	Page
Tests for safety	26	20
Third step (trigger squeeze)	48	40
Training horses	72	71
Trigger squeeze (third step)	48	40
Unload pistol	34, 43	23, 25
Withdraw magazine	29, 39	22, 24

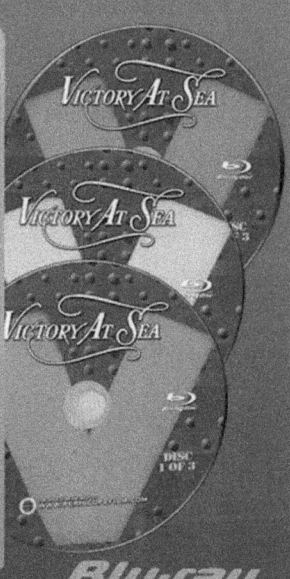

©2013 Periscope Film LLC
All Rights Reserved
ISBN#978-1-940453-04-0
www.PeriscopeFilm.com

www.ingramcontent.com/pod-product-compliance
Lightning Source LLC
Chambersburg PA
CBHW072201100426

42738CB00011BA/2495